New 12^95

Engineer-In-Training License Review

Ninth Edition

Donald G. Newnan
C. Dean Newnan
School of Engineering
San Jose State University

Engineering Press, Inc. San Jose, California 95103-0001

Library of Congress Cataloging in Publication Data

Newnan, C Dean.
Engineer-in-training license review.

Includes index.
1. Engineering--Examinations, questions, etc.
I. Title.
TA159.N39 1981 620'.0076 80-22287
ISBN 0-910554-33-1

Printed in the United States of America

Available to booksellers from
Medical and Technical Books, Inc.
11511 Tennessee Avenue, Los Angeles, CA 90064
and to Member Stores from
NACSCORP, Inc.
528 East Lorain Street, Oberlin, OH 44074
or from the publisher
Engineering Press, Inc.
P.O. Box 1 San Jose, CA 95103-0001

Contents

	Preface	v
1.	Mathematics and Mathematical Modeling	1
2.	Statics	17
3.	Dynamics	33
4.	Mechanics of Materials	49
5.	Materials Science	87
6.	Fluid Mechanics	103
7.	Thermodynamics	137
8.	Electrical Circuits	187
9.	Engineering Economics	207
10.	Chemistry	235
11.	Structure of Matter	241
12.	Computer Programming	253
	Index	263

ENGINEERING LICENSE REVIEW BOOKS

Engineering Press publishes books to help engineers prepare for the professional registration examinations. We have a mail order department to sell our books, and others, directly to individuals and groups by mail.

Some of the books are:

Engineer-In-Training License Review
Preparing for the Engineer-In-Training Examination
Civil Engineering License Review
Preparing for the Civil Engineering Professional Examination
Sanitary Engineering Problems for the Professional Engineer
Mechanical Engineering License Review
Electrical Engineering License Review
Economic Analysis for the Professional Engineer Examination

For a free copy of our complete catalog of available books, with descriptions, prices, and ordering information, write to

Engineering Press, Inc.
P. O. Box 1
San Jose, California 95103

Preface

Registration as a Professional Engineer is achieved by passing an Engineer-In-Training or Engineering Fundamentals examination, followed by an exam in a specific branch of engineering. The first examination, whether called Engineer-In-Training, Intern Engineer, or Engineering Fundamentals, is a test of the applicant's knowledge of fundamentals of engineering and science. The second or professional engineering examination is a test of the applicant's ability to apply engineering principles and judgement to professional problems. Here we are concerned exclusively with the Engineer-In-Training or Engineering Fundamentals examination.

The examination consists of two four-hour periods. The morning session of four hours presently contains 140 multiple choice questions. The emphasis is:

Comprehension and knowledge	2/3
Evaluation, analysis, and application	1/3

There are five choices from which to select the correct answer for each question. The subjects and approximate number of questions are:

Mathematics	12
Mathematical Modeling of Engineering Systems	8
Statics	13
Dynamics	12
Mechanics of Materials	13
Materials Science	6
Fluid Mechanics	14
Thermodynamics	14
Electrical Circuits	18
Engineering Economics	6
Chemistry	10
Structure of Matter	6
Computer Programming	8
	140

Questions are bunched by groups in the above categories. The applicant is required to answer all of the 140 questions, and all will be machine graded to determine a numerical score for the morning session.

It is suggested that you carefully pace your time in the morning, for an average of only 1.7 minutes per question may permit understanding the questions, but allows very little time for computation, and permits essentially no time to consult reference materials. One strategy used by many applicants has been to rapidly work through the entire group of questions, answering the easy, fast ones, then to return to the more demanding questions in areas of personal competence. Finally, to guess as necessary to complete the questions. Any well designed examination of this type has correct answers randomly distributed among the five choices, averaging 20% in each choice.

In the afternoon four-hour session all applicants are required to solve 50 questions as follows:

Subject	Questions
Engineering Mechanics (Statics and Dynamics)	15
Mathematics	15
Electrical Circuits	10
Engineering Economics	10
	50

In addition, each applicant is required to solve 20 problems in two of five additional subject areas:

Subject	Questions
Computer Programming	10
Electronics and Electrical Machinery	10
Fluid Mechanics	10
Mechanics of Materials	10
Thermodynamics/Heat Transfer	10
	50

Thus the applicant must do 70 problems in the afternoon. The problems will require the application of engineering principles to more complex problems in the afternoon than in the morning.

The test is generally open book, but there is considerable variation among States regarding what may be brought into the examination. Some permit an unlimited amount of books, notes, and so on. Others permit only standard reference materials like handbooks and textbooks. Still others limit the total quantity of reference material brought into the examination.

The over 450 problems contained in this edition have been selected to offer guidance in your systematic review of the 12 subject areas. I have appreciated past reader comments, and would appreciate being told about errors that are noted.

D.G.N.

1

Mathematics and Mathematical Modeling

MATH 1

The integral of $y = x^3 - x + 1$ is:

(a) $3x^2 - 1 + C$

(b) $\frac{x^3}{3} - \frac{x^2}{2} + x$

(c) $\frac{x^4}{3} - \frac{x^2}{2} + 1 + C$

(d) $\frac{x^4}{3} - \frac{x^2}{2} + 1$

(e) $\frac{x^4}{4} - \frac{x^2}{2} + x + C$

$$\int y\,dx = \int (x^3 - x + 1)dx = \frac{x^4}{4} - \frac{x^2}{2} + x + C$$

Answer is (e)

MATH 2

The only point of inflection on the curve representing the equation $y = x^3 + x^2 - 3$ is at x equals:

(a) -2/3
(b) -1/3
(c) 0
(d) 1/3
(e) 2/3

Maxima, minima and points of inflection:

Take first derivative, equate it to zero, and solve for its roots.

Take second derivative, insert the above roots to determine whether each is a maximum or a minimum.

Equate second derivative to zero and solve for its root(s) or point(s) of inflection.

	Max.	Min.	Inflection
$\frac{dy}{dx}$	0	0	anything
$\frac{d^2y}{dx^2}$	-	+	0

Apply the above to the stated problem: $y = x^3 + x^2 - 3$

$\frac{dy}{dx} = 3x^2 + 2x = 0 \qquad x(3x + 2) = 0 \qquad \therefore x = 0, -2/3$
(max. and min. exist here)

$\frac{d^2y}{dx^2} = 6x + 2 \qquad$ at $x = 0$: $\frac{d^2y}{dx^2} = +2$, minimum exists

at $x = -2/3$: $\frac{d^2y}{dx^2} = -4 + 2 = -2$, maximum exists

$\frac{d^2y}{dx^2} = 6x + 2 = 0 \qquad x = -1/3$, inflection point exists ●

Answer is (b)

MATH 3

There are 10 defectives per 1,000 items of a product in the long run. What is the probability that there is one and only one defective in a random lot of 100?

(a) 99×0.01^{99}
(b) 0.01
(c) 0.5
(d) 0.99^{100}
(e) 0.99^{99}

The problem involves the binomial probability.

$P_{defective} = \frac{10}{1000} = 0.01$ This is the probability that one item selected at random is defective.

$P_{good} = 1 - P_{defective} = 0.99$

The probability that exactly one defective will be found in a random sample of 100 items is given by the binomial: b(1, 100, 0.01).

$b(1, 100, 0.01) = {}_{100}C_1 \, (0.01)^1 \, (0.99)^{99}$

(annotations: no. defective → 1 and exponent 1; size sample → 100; $P_{defective}$ → 0.01; P_{good} → 0.99; no. good → 99)

$_nC_r = \binom{n}{r} = \frac{n!}{(n-r)!\,r!}$ = number of combinations of n objects taken r at a time without concern for order of arrangement.

$_{100}C_1 = \frac{100!}{99!\,1!} = 100 \qquad b(1, 100, 0.01) = 100(0.01)(0.99)^{99} = 0.99^{99}$ ●
$= 0.3697$

Answer is (e)

MATH 4

If the second derivative of the equation of a curve is equal to the negative of the equation of that same curve, the curve is

(a) an exponential
(b) a tangent
(c) a conic section
(d) a sinusoid
(e) a parabola

Taking the sine curve

$f(x) = \sin x \qquad f'(x) = \cos x \qquad f''(x) = -\sin x$

Thus a function whose second derivative is equal to the negative of the equation of that same function is a sinusoid. ●

Answer is (d)

MATH 5

To find the angles of a triangle, given only the lengths of the sides, one would use

(a) the law of cosines
(b) the law of tangents
(c) the law of sines
(d) the inverse-square law
(e) orthogonal functions

The law of cosines is

$$a^2 = b^2 + c^2 - 2bc \cos A$$

for any plane triangle with angles A, B, C and sides a, b, c respectively.

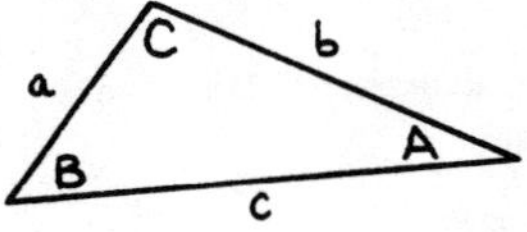

This law can be applied to solve for the angles, given three sides in a plane triangle.

Answer is (a) ●

MATH 6

The rationalized value of the complex number $\frac{6 + i\,2.5}{3 + i\,4}$ is

(a) $2 + i\,0.625$

(b) $\left(\frac{6 + i\,2.5}{3 + i\,4}\right)\left(\frac{6 - i\,2.5}{3 + i\,4}\right)$

(c) $\left(\frac{6 + i\,2.5}{3 + i\,4}\right)\left(\frac{6 + i\,2.5}{3 + i\,4}\right)$

(d) $\frac{28 - i\,16.5}{25}$

(e) $1 - i$

To rationalize, multiply the numerator and denominator of the complex fraction by the complex conjugate of the denominator.

$$\frac{6 + i\,2.5}{3 + i\,4} \times \frac{3 - i\,4}{3 - i\,4} = \frac{18 - i\,24 + i\,7.5 + 10}{9 - i\,12 + i\,12 + 16} = \frac{28 - i\,16.5}{25}$$ ●

Answer is (d)

MATH 7

Study the series of numbers to discover the "SYSTEM" in which they are arranged.

For the series 1 5 14 30 __ 91 the fifth term is

(a) 59
(b) 53
(c) 61
(d) 50
(e) 55

If we are given any series of values it is always possible to find a polynomial which will pass through the given points. So the equation of the polynomial would be the "system." Thus any number between 30 and 91 could be used.

	1		5		14		30		x		91
first differences:		4		9		16		y		z	
		2^2		3^2		4^2		5^2		6^2	

We see that the first differences are successive perfect squares.

Thus $x - 30 = y = 5^2 = 25$ $\qquad x = 30 + 25 = 55$ ●

Answer is (e)

MATH 8

If $\sin\alpha = \dfrac{a}{\sqrt{a^2 + b^2}}$ which of the following statements is true?

(a) $\tan^{-1}\dfrac{b}{a} = \dfrac{\pi}{2} - \alpha$

(b) $\tan^{-1}\dfrac{b}{a} = -\alpha$

(c) $\cos^{-1}\dfrac{b}{\sqrt{a^2 + b^2}} = \dfrac{\pi}{2} - \alpha$

(d) $\cos^{-1}\dfrac{a}{\sqrt{a^2 + b^2}} = \alpha$

(e) None of the above

So the triangle looks like this:

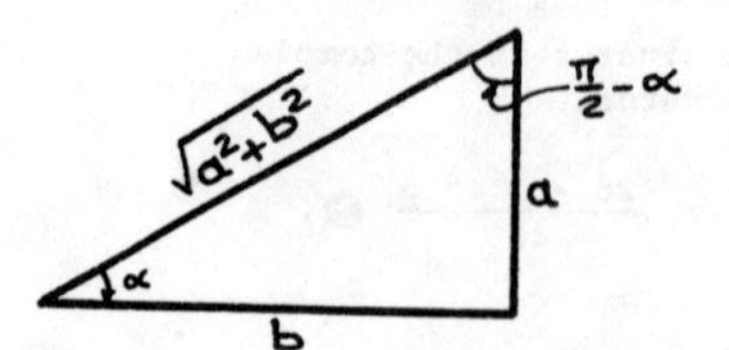

$\tan\left(\dfrac{\pi}{2} - \alpha\right) = \dfrac{b}{a}$

$\therefore \tan^{-1}\dfrac{b}{a} = \dfrac{\pi}{2} - \alpha$ ●

Answer is (a)

MATH 9

In finding the distance between two points P_1 (X_1, Y_1) and P_2 (X_2, Y_2), the most direct procedure is to use:

(a) The translation of the axes
(b) The Pythagorean Theorem
(c) The slope of the line
(d) The derivative
(e) The integral

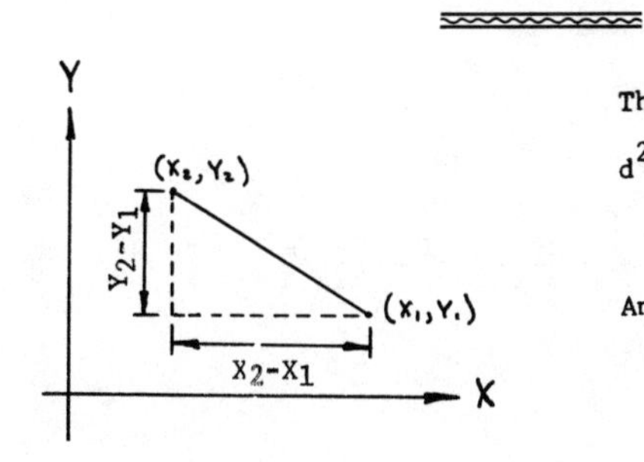

The Pythagorean Theorem

$$d^2 = (X_2 - X_1)^2 + (Y_2 - Y_1)^2$$

Answer is (b) ●

MATH 10

The sine of $840°$ equals:

(a) $-\cos 30°$
(b) $-\cos 60°$
(c) $\sin 30°$
(d) $\sin 60°$
(e) $-\sin 30°$

$$840° = 2(360) + 120 = 2\pi + 120°$$

$$\sin (2\pi + 120°) = \sin 120° = \sin (180° - 60°) = \sin 60°$$ ●

Answer is (d)

MATH 11

One root of $x^3 - 8x - 3 = 0$ is:

(a) 2
(b) 3
(c) 4
(d) 5
(e) 6

The solution is obtained by seeing which of the 5 answers satisfies the equation.

x	x^3-8x-3
2	-11
3	0 ●
4	29
5	82
6	165

Answer is (b)

MATH 12

The value of Tan (A + B), where Tan A = $\frac{1}{3}$ and Tan B = $\frac{1}{4}$ is (Note: A and B are acute angles)

(a) 7/12
(b) 1/11
(c) 7/11
(d) 7/13
(e) none of the above

$$\text{Sin}(A + B) = \text{Sin } A \text{ Cos } B + \text{Cos } A \text{ Sin } B$$

$$\text{Cos}(A + B) = \text{Cos } A \text{ Cos } B - \text{Sin } A \text{ Sin } B$$

$$\text{So Tan}(A + B) = \frac{\text{Sin}(A + B)}{\text{Cos}(A + B)} = \frac{\text{Sin } A \text{ Cos } B + \text{Cos } A \text{ Sin } B}{\text{Cos } A \text{ Cos } B - \text{Sin } A \text{ Sin } B}$$

dividing by Cos A Cos B

$$\text{Tan}(A + B) = \frac{\frac{\text{Sin } A \text{ Cos } B}{\text{Cos } A \text{ Cos } B} + \frac{\text{Cos } A \text{ Sin } B}{\text{Cos } A \text{ Cos } B}}{\frac{\text{Cos } A \text{ Cos } B}{\text{Cos } A \text{ Cos } B} - \frac{\text{Sin } A \text{ Sin } B}{\text{Cos } A \text{ Cos } B}}$$

$$= \frac{\text{Tan } A + \text{Tan } B}{1 - \text{Tan } A \text{ Tan } B}$$

$$= \frac{\frac{1}{3} + \frac{1}{4}}{1 - \frac{1}{3} \times \frac{1}{4}} = \frac{\frac{4}{12} + \frac{3}{12}}{1 - \frac{1}{12}} = \frac{\frac{7}{12}}{\frac{11}{12}} = \frac{7}{11}$$ ●

Note that the problem could also be solved by determining Angle A (whose tangent is 1/3) and Angle B (whose tangent is 1/4). Then we could find the tangent of (A + B).

$\text{Tan}^{-1} \frac{1}{3} = 18.435°$ $\text{Tan}^{-1} \frac{1}{4} = 14.036°$

$\text{Tan}(18.435° + 14.036°) = \text{Tan}(32.471°) = 0.6364 = \frac{7}{11}$ ●

Answer is (c)

MATH 13

To cut a right circular cone in such a was as to reveal a parabola, it must be cut

(a) perpendicular to the axis of symmetry
(b) at any acute angle to the axis of symmetry
(c) at any obtuse angle to the axis of symmetry
(d) parallel to the axis of symmetry
(e) none of these

To reveal a parabola, a right circular cone must be cut parallel to an element of the cone and intersecting the axis of symmetry. ●

Answer is (e)

MATH 14

Naperian logarithms have a base of:

(a) 3.1416
(b) 2.171828
(c) 10
(d) Any positive number can be used
(e) 2.71828

Common logarithms have base 10.
Naperian or natural logarithms have base e = 2.71828 ●

Answer is (e)

MATH 15

$(5.743)^{1/30}$ equals:

(a) 1.03
(b) 1.04
(c) 1.05
(d) 1.06
(e) 1.07

$$\text{Log } (5.743)^{1/30} = \frac{1}{30}\text{ Log } 5.743 = \frac{1}{30}(0.7592) = 0.0253$$

The antilogarithm of 0.0253 is 1.06 ●

Answer is (d)

MATH 16

For a given curve f(x,y) = 0, it is found that f(x,y) = f(x,-y). This means that

(a) The curve is unsymmetrical
(b) The curve is a conic section
(c) The curve is symmetrical to both axes
(d) The curve is symmetrical to the x-axis
(e) The curve is symmetrical to the y-axis

Answer is (d) ●

MATH 17

The curve in the figure below has the equation y = f(x). At point A, what are the values of $\frac{dy}{dx}$ and $\frac{d^2y}{dx^2}$?

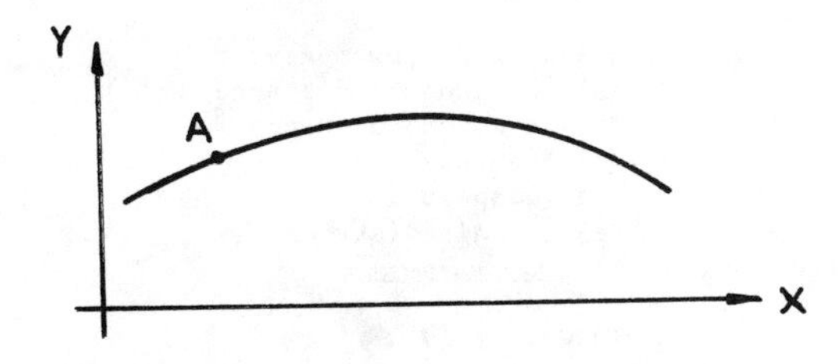

(a) $\frac{dy}{dx} < 0, \quad \frac{d^2y}{dx^2} < 0$

(b) $\frac{dy}{dx} < 0, \quad \frac{d^2y}{dx^2} > 0$

(c) $\frac{dy}{dx} = 0, \quad \frac{d^2y}{dx^2} = 0$

(d) $\frac{dy}{dx} > 0, \quad \frac{d^2y}{dx^2} > 0$

(e) $\frac{dy}{dx} > 0, \quad \frac{d^2y}{dx^2} < 0$

The first derivative $\frac{dy}{dx}$ is the slope of the curve. At point A the slope is positive. The second derivative $\frac{d^2y}{dx^2}$ gives the direction of bending. A negative value indicates the curve is concave downward.

Answer is (e) ●

MATH 18

The expression $\frac{6!}{3!\ 0!}$ is equal to

(a) ∞
(b) 120
(c) 2!
(d) 0
(e) 3!

$$\frac{6!}{3!\ 0!} = \frac{6 \times 5 \times 4 \times \cancel{3!}}{\cancel{3!} \times 1} = 120$$ ●

Answer is (b)

MATH 19

$$\lim_{x \to 1} \frac{x^2 - 1}{x - 1} =$$

(a) 2
(b) ∞
(c) 0
(d) 1
(e) none of these

$$\lim_{x \to 1} \frac{x^2 - 1}{x - 1} = \lim_{x \to 1} \frac{\cancel{(x - 1)}(x + 1)}{\cancel{(x - 1)}} = \lim_{x \to 1} (x + 1) = 2$$ ●

Answer is (a)

MATH 20

In statistics, the standard deviation measures

(a) a standard distance
(b) a normal distance
(c) central tendency
(d) dispersion
(e) a unit distance

Answer is (d) ●

MATH 21

There are 3 bins containing integrated circuits. One bin has 2 premium IC's, one has 2 regular IC's, and one has 1 premium IC and 1 regular IC.

You choose an IC at random without looking into any of the bins. The chosen IC is found to be a premium IC.

What is the probability that the remaining IC in that bin is also a premium IC?

(a) 1/5
(b) 1/4
(c) 1/3
(d) 2/3
(e) 1/2

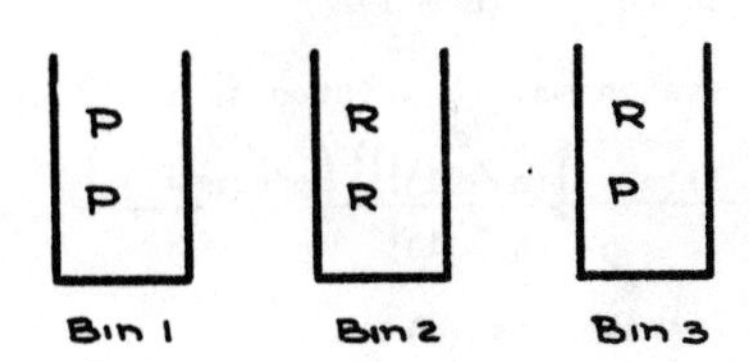

Since the first IC chosen is a premium IC, it was drawn from either Bin 1 or Bin 3. From the distribution of premium IC's, the probability that the premium IC came from Bin 1 is 2/3, and from Bin 3 is 1/3.

In Bin 1 the probability that the remaining IC is a premium IC is 1; in Bin 3 the probability is 0. Thus the probability that the remaining IC is a premium IC is:

$$\frac{2}{3}(1) + \frac{1}{3}(0) = \frac{2}{3}$$ ●

Alternate solution using Bayes' Theorem for conditional probability:

$$P(\text{Bin 1} \mid \text{Drew Premium}) = \frac{P(\text{Bin 1 and Premium})}{P(\text{Premium})} = \frac{P(\text{Premium} \mid \text{Bin 1}) \cdot P(\text{Bin 1})}{\sum_{i=1}^{3} P(\text{Premium} \mid \text{Bin 1}) \cdot P(\text{Bin 1})}$$

$$= \frac{1(\frac{1}{3})}{1(\frac{1}{3}) + 0(\frac{1}{3}) + \frac{1}{2}(\frac{1}{3})} = \frac{2}{3}$$ ●

Answer is (d)

MATH 22

If a right circular cone is cut parallel with the axis of symmetry, you would reveal

(a) circle
(b) hyperbola
(c) eclipse
(d) parabola
(e) none of these

Answer is (b) ●

MATH 23

The simplest value of $\frac{[(n+1)!]^2}{n!\,(n-1)!}$ is

(a) n^2
(b) $n(n + 1)$
(c) $n + 1$
(d) $n(n + 1)^2$
(e) $\frac{n+1}{n-1}$

The value $(n + 1)!$ may be written as $(n + 1)(n)\,[(n - 1)!]$

It may also be written as $n!\,(n + 1)$

Hence the given expression may be written as follows:

$$\frac{\{(n+1)(n)\,[(n-1)!]\}\{n!\,(n+1)\}}{n!\,(n-1)!} = (n+1)^2 n$$

Answer is (d)

MATH 24

If $X^{3/4} = 8$, X equals

(a) 6
(b) 9
(c) -9
(d) 16
(e) 20

Raise both sides of the equation to the 4/3 power

$$\left[X^{3/4}\right]^{4/3} = 8^{4/3}$$

$$X = \sqrt[3]{8^4} = \sqrt[3]{(2^3)^4} = 2^{\frac{3 \cdot 4}{3}} = 2^4 = 16$$

Answer is (d)

MATH 25

If $\log_a 10 = 0.250$, $\log_{10} a$ equals

(a) 4
(b) .50
(c) 2
(d) .25
(e) 1,000

$\log_a 10 = 0.250$ can be written as $10 = a^{0.250}$

taking $\log_{10}$ $\qquad \log_{10} 10 = \log_{10} a^{0.250}$

$$1 = 0.250 \log_{10} a$$

Since $1 = 0.250 \log_{10} a$

$$\log_{10} a = \frac{1}{0.250} = 4$$

Answer is (a)

MATH 26

The csc of 960° is equal to

(a) $-\frac{2\sqrt{3}}{3}$
(b) 1
(c) 1/2
(d) -2
(e) $-\frac{\sqrt{3}}{2}$

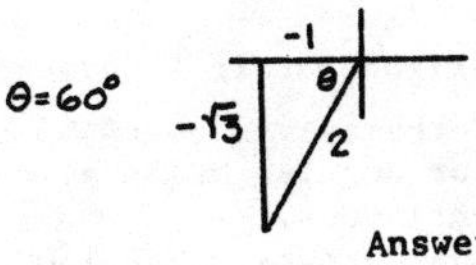

$$\csc 960° = \csc(5\pi + 60)$$

$$= \csc\theta = \frac{r}{y} = \frac{2}{-\sqrt{3}} = -\frac{2\sqrt{3}}{3}$$

Answer is (a)

MATH 27

If $i = \sqrt{-1}$, the quantity i^{27} is equal to

(a) 0
(b) i
(c) -i
(d) 1
(e) -1

$$i^{27} = (\sqrt{-1})^{27} = \sqrt{(-1)^{27}} = \sqrt{(-1)^{26}(-1)} = (-1)^{13}\sqrt{-1} = -1\sqrt{-1}$$

now substituting back i for $\sqrt{-1}$

$$i^{27} = -1(i) = -i$$

Answer is (c)

MATH 28

If the first derivative of the equation of a curve is a constant, the curve is a

(a) circle
(b) hyperbola
(c) parabola
(d) straight line
(e) sine wave

If $\frac{dy}{dx} = m, \quad y = \int m\,dx = m\int dx = mx + b$

so $y = mx + b$ a straight line

Answer is (d)

MATH 29

An angle between 90° and 180° has:

(a) A positive sine and cosine
(b) A negative cotangent and cosecant
(c) A negative secant and tangent
(d) All its trigonometric functions negative
(e) All its trigonometric functions positive

In the second quadrant all functions are negative except the sine and cosecant.

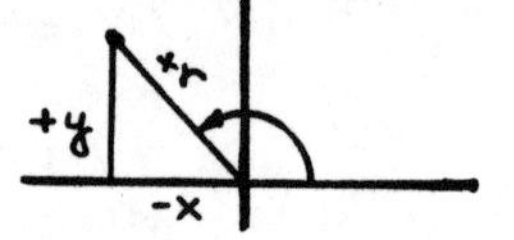

Answer is (c)

MATH 30

The following is a characteristic of all trigonometric functions:

(a) The values of all functions repeat themselves every 45 degrees.
(b) All functions are units of length or angular measure.
(c) The graphs of all functions are continuous.
(d) The values of all functions are never greater than 1.00.
(e) All functions are dimensionless units.

All functions are ratios of lengths with the result that they are dimensionless.

Answer is (e)

MATH 31

Given the function $P = 2R^2S^3T^{1/2} + R^{1/3}S \sin 2T$

$\frac{\partial P}{\partial T}$ is

(a) $R^2S^3T^{1/2} + 2R^{1/3}S \cos 2T$

(b) $R^2S^3T^{-1/2} + 2R^{1/3}S \cos 2T$

(c) $2R^2S^3T^{-1/2} + R^{1/3}S \cos 2T$

(d) $R^2S^3T^{-1/2} + 2R^{1/3}S \sin 2T$

(e) $2R^2S^3T^{1/2} + R^{1/3}S \sin 2T$

In finding $\frac{\partial P}{\partial T}$ R and S are treated as constants.

$P = 2R^2S^3T^{1/2} + R^{1/3}S \sin 2T$

$\frac{\partial P}{\partial T} = 2R^2S^3(\frac{1}{2}T^{-1/2}) + R^{1/3}S(\cos 2T)(2)$

$= R^2S^3T^{-1/2} + 2R^{1/3}S \cos 2T$

Answer is (b)

MATH 32

For a given curve $y = f(x)$ that is continuous between $x = a$ and $x = b$, the average value of the curve between the ordinates at $x = a$ and $x = b$ is represented by:

(a) $\dfrac{\int_a^b x^2\,dy}{(b-a)}$ (b) $\dfrac{\int_a^b y^2\,dx}{(b-a)}$ (c) $\dfrac{\int_a^b x\,dy}{(a-b)}$

(d) $\dfrac{\int_a^b y\,dx}{(a-b)}$ (e) $\dfrac{\int_a^b y\,dx}{(b-a)}$

$$\text{Area} = \int_a^b y\,dx$$

$$\text{Average value} = \frac{\text{area}}{\text{base width}} = \frac{\int_a^b y\,dx}{b-a}$$

Answer is (e)

MATH 33

The probability that both stages of a 2-stage missile will function correctly is 0.95. The probability that the first stage will function correctly is 0.98.

What is the probability that the second stage will function correctly given that the first one does?

(a) 0.99
(b) 0.98
(c) 0.97
(d) 0.95
(e) 0.93

Given: $P(S_1) = 0.98$ $P(S_2 \cap S_1) = 0.95$ Find: $P(S_2 | S_1)$

$$P(S_2 | S_1) = \frac{P(S_2 \cap S_1)}{P(S_1)} = \frac{0.95}{0.98} = 0.97$$

Answer is (c)

MATH 34

If $y = \cos x$, $\dfrac{dy}{dx} = ?$

(a) $\sin x$
(b) $-\tan x \cos x$
(c) $\dfrac{1}{\sec x}$
(d) $\sec x \sin x$
(e) $-\dfrac{1}{\cos x}$

$\frac{dy}{dx} = -\sin x$ Since $\tan x = \frac{\sin x}{\cos x}$, then $\sin x = \tan x \cos x$

Thus we can define the derivative

$\frac{dy}{dx} = -\tan x \cos x$ ●

Answer is (b)

MATH 35

The value of the determinant $\begin{vmatrix} 1 & 1 & 0 & 0 \\ 1 & 1 & 1 & 0 \\ 0 & 1 & 1 & 1 \\ 0 & 0 & 1 & 1 \end{vmatrix}$ is

(a) +1
(b) 0
(c) -2
(d) -1
(e) none of the above values

$$\begin{vmatrix} 1 & 1 & 0 & 0 \\ 1 & 1 & 1 & 0 \\ 0 & 1 & 1 & 1 \\ 0 & 0 & 1 & 1 \end{vmatrix} = 1 \begin{vmatrix} 1 & 1 & 0 \\ 1 & 1 & 1 \\ 0 & 1 & 1 \end{vmatrix} - 1 \begin{vmatrix} 1 & 0 & 0 \\ 1 & 1 & 1 \\ 0 & 1 & 1 \end{vmatrix}$$

$$\cancel{= 1 \begin{vmatrix} 1 & 1 \\ 1 & 1 \end{vmatrix}} - 1 \begin{vmatrix} 1 & 0 \\ 1 & 1 \end{vmatrix} \cancel{-1 \begin{vmatrix} 1 & 1 \\ 1 & 1 \end{vmatrix}} + 1 \begin{vmatrix} 0 & 0 \\ 1 & 1 \end{vmatrix} = -1(1) + 1(0) = -1$$ ●

Answer is (d)

MATH 36

In matrix notation:

$$\begin{bmatrix} 3 & 7 \\ 2 & 6 \end{bmatrix} \begin{bmatrix} X_1 \\ X_2 \end{bmatrix} = \begin{bmatrix} 2 \\ 4 \end{bmatrix}$$

Find X_1 and X_2

$3X_1 + 7X_2 = 2$ (1)

$2X_1 + 6X_2 = 4$ (2)

$2 \times (1)$ $6X_1 + 14X_2 = 4$

$-3 \times (2)$ $-6X_1 - 18X_2 = -12$

$-4X_2 = -8$ $X_2 = 2$ ●

$3X_1 + 7(2) = 2$ $X_1 = -4$ ●

MATH 37

The step response of a unity feedback control system has a high percentage overshoot. One can then assume the system to be

(a) Overdamped.
(b) Unstable.
(c) To be at least a second order system.
(d) To be a first order system.
(e) To have a high amount of system error.

Since the response to the step input has a high overshoot, the system itself must be described by at least a second order differential equation. A first order differential equation solution has no sinusoidal term in it, and therefore cannot exceed it's final value. Also, since to have a high amount of overshoot, the system's gain must be relatively high, therefore the system error is relatively small. Thus answer (c) is the only possible choice. ●

Answer is (c)

MATH 38

A Bode plot of a transfer function has a slope for the lowest frequency asymptote of -6 db/octive. One can then assume that

(a) The transfer function this represents may be characterized by at least one integration.
(b) The transfer function this represents may be characterized by two or more pure integrations.
(c) If the system gain is low, the system (for a unity feedback system) is highly unstable.
(d) The corresponding Nyquist diagram starts at -180° (for the lowest frequency).
(e) The corresponding root locus diagram has no pole at the origin.

Because the lowest frequency asymptote has a -6 db/octive slope (this is equivalent to -20 db/decade), the form of the transfer function must be

$$G(j\omega) = \frac{K(?)}{S(?)(?)}$$

Thus the single "S" in the denominator may be represented by one pure integration in the time domain. Since, for answer (c) the gain is low, the closed loop system is very slowly responding and the phase shift would be near -90° at the gain crossover point. Thus a unity feedback system would more than likely be far from the -1 point, resulting in a highly stable system. Thus only answer (a) is correct. ●

Answer is (a)

MATH 39

For a unity feedback control system, the system error is finite if

(a) The input is a step and the forward transfer function has at least one (pure) integration.
(b) The input is a step and the forward transfer function has at least two integrations.

(c) The input is a ramp (i.e., constant velocity input) and the forward transfer function has one (pure) integration.
(d) The input is a ramp and the transfer function has two (pure) integrations.
(e) The forward transfer function has all zeros and no poles.

System error for any particular input may easily be found by use of final value theorem and considering that the error, E(s), is given by

$$E(s) = \left[\frac{1}{1 + G(s)}\right] R(s)$$

Then one (mentally) can set up the final value theorem to give error

$$e(t)\Big|_{t\to\infty} = \lim_{s\to 0} s\,E(s) \longrightarrow \text{answer.}$$

Thus, since for a ramp

$$R(s) = \frac{1}{S^2}$$

answer (c) will be the only answer to give a finite error. ●

Answer is (c)

MATH 40

An armature controlled d.c. shunt control motor (with fixed field) may be represented by

(a) $G(S) = \dfrac{K}{S^2(S\tau + 1)} = \dfrac{\Theta(S)}{E(S)}$

(b) $G(S) = \dfrac{SK}{(S\tau + 1)} = \dfrac{\Theta(S)}{E(S)}$

(c) The same approximate transfer function as a two energy storage device (i.e., a spring, mass, and damper system).
(d) The same approximate transfer function as a two-phase a.c. control motor (i.e., with one field controlled and the other to a reference source).
(e) The same approximate transfer function as a frictional only system (i.e., only damping).

The one transfer function that is most familiar (and having derived several different ways) is the control motor (either d.c. or a.c.). This transfer function comes out to be of the form of

$$G(S) = \frac{K}{S(S\tau + 1)}$$

and should be in one's own "memory bank." Thus answer (d) is the only applicable solution. ●

Answer is (d)

2

Statics

STATICS 1

The magnitude of the force acting in the member BC of the truss shown below is:

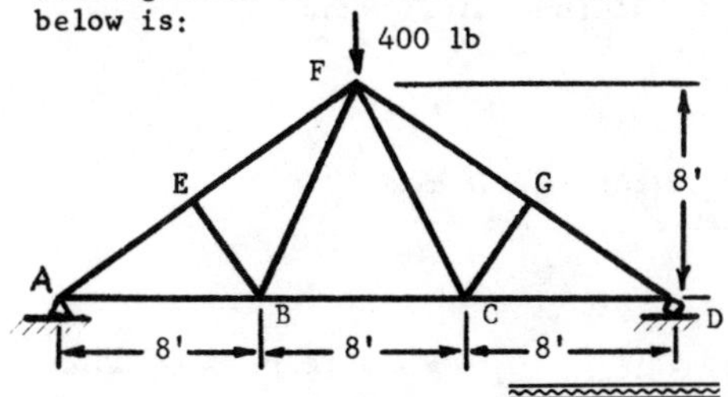

(a) 200 lb.
(b) 300 lb.
(c) 360 lb.
(d) 400 lb.
(e) 480 lb.

By symmetry determine the support reaction A_y to be 200 lb.

Although A is pinned, there are no horizontal forces on the truss, so $A_x = 0$.

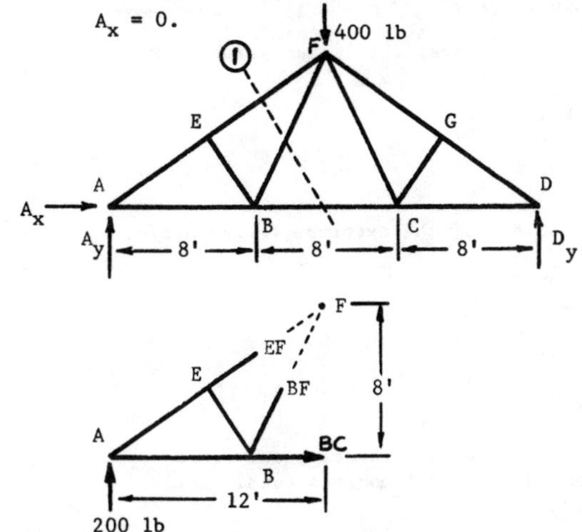

Section the truss along (1) to cut BC. Draw the free body diagram of the structure to the left of the section.

Take moments about F. (Counterclockwise is positive.)

$$M_F = 0 = -12(200) + 8(BC)$$

BC = +300 lb ●
Tension

Answer is (b)

STATICS 2

The coefficient of friction between the 60-pound block and the plane shown is 0.30. If the block is to remain in equilibrium, what is the maximum allowable magnitude for the force P?

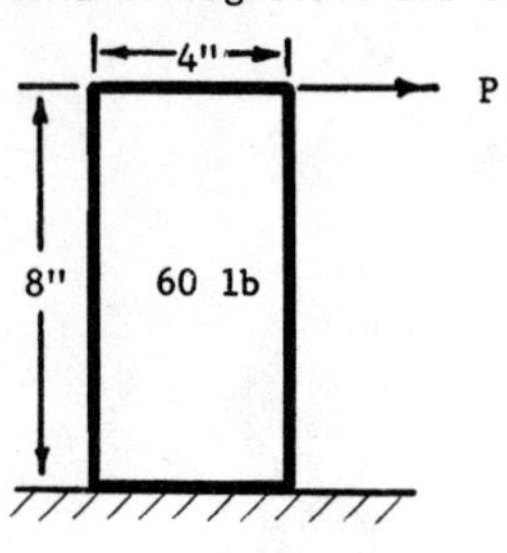

(a) 12 lb.
(b) 15 lb.
(c) 18 lb.
(d) 21 lb.
(e) 24 lb.

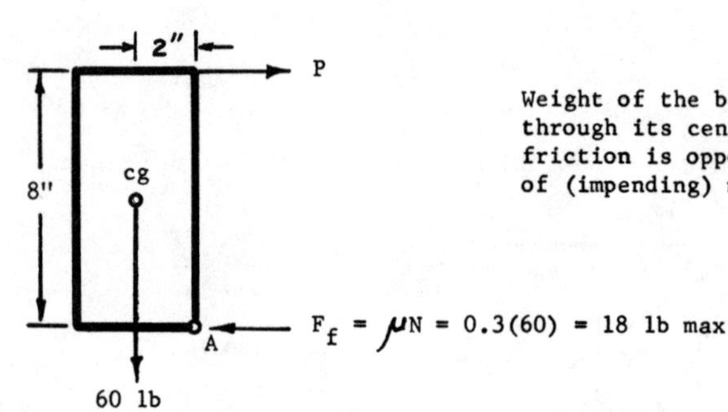

Weight of the block acts downward through its center of gravity and friction is opposite to direction of (impending) motion.

$F_f = \mu N = 0.3(60) = 18$ lb max

To avoid translation: $F_x = 0 = +P - 18_{max}$ $\therefore$ P = + 18 lb. max. →

To avoid tipping about A:

$M_A = 0 = -8P + 2(60)$ $\therefore$ P = + 15 lb. max. →

Choose the lesser value of P, or 15 lb. ●

Answer is (b)

STATICS 3

For a system to be in equilibrium, the sum of the external forces acting on the system must be

(a) equal to unity
(b) a maximum
(c) indeterminant
(d) zero
(e) infinite

Conditions of equilibrium for a system are:

1. The sum of the external forces equal zero
2. The sum of the external moments equal zero

Answer is (d) ●

STATICS 4

Derive the formula for computing the moment of interia of an equilateral triangle about one of its sides.

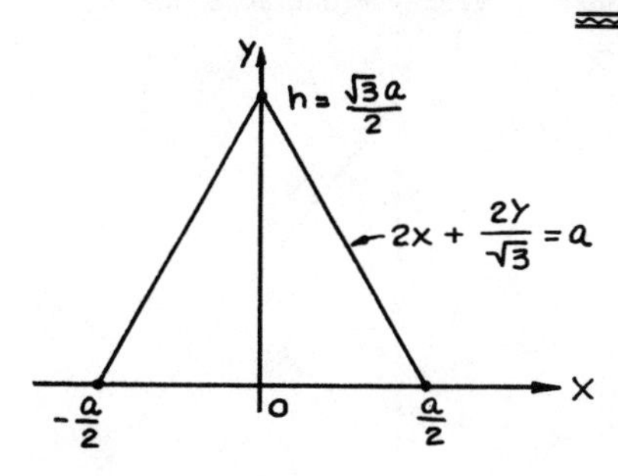

Altitude h

$$\left(\frac{a}{2}\right)^2 + h^2 = a^2 \qquad h^2 = a^2 - \frac{a^2}{4} = \frac{3a^2}{4}$$

$$h = \frac{\sqrt{3}\,a}{2}$$

Equation of side in terms of x and y

eqn of straight line (intercept form)

$$\frac{x}{\frac{a}{2}} + \frac{y}{\frac{\sqrt{3}\,a}{2}} = 1$$

$$2x + \frac{2y}{\sqrt{3}} = a \qquad 2x = a - \frac{2y}{\sqrt{3}}$$

Moment of Interia

$$I_x = \int y^2\,dA = \int y^2\left(a - \frac{2}{\sqrt{3}}\,y\right) dy = \int_0^{\frac{\sqrt{3}a}{2}} \left[ay^2 - \frac{2}{\sqrt{3}}\,y^3\right] dy$$

$$= \left[\frac{ay^3}{3} - \frac{2y^4}{4\sqrt{3}}\right]_0^{\frac{\sqrt{3}a}{2}}$$

$$= \frac{\sqrt{3}\,a^4}{8} - \frac{3\sqrt{3}\,a^4}{32} = \frac{\sqrt{3}}{32}\,a^4$$ ●

STATICS 5

A tripod whose legs are each 10 feet long supports a load of 1000 pounds.
The feet of the tripod are at the vertices of a horizontal isosceles triangle whose base is 12 feet and whose altitude is 8 feet.
Determine the total load in each leg.

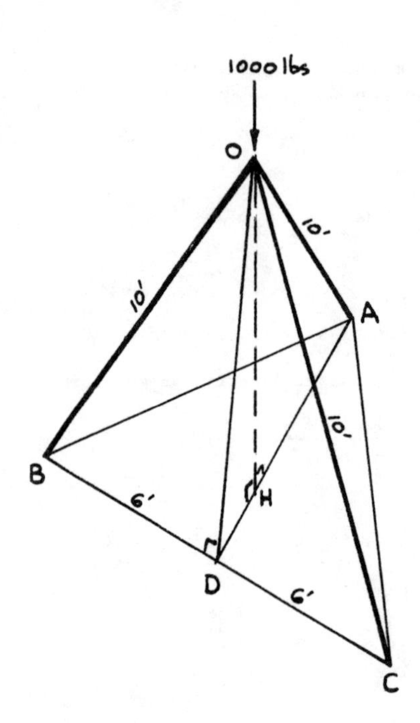

From the figure

$$OD = \sqrt{10^2 - 6^2} = \sqrt{64} = 8$$

Consider triangle ODA separately

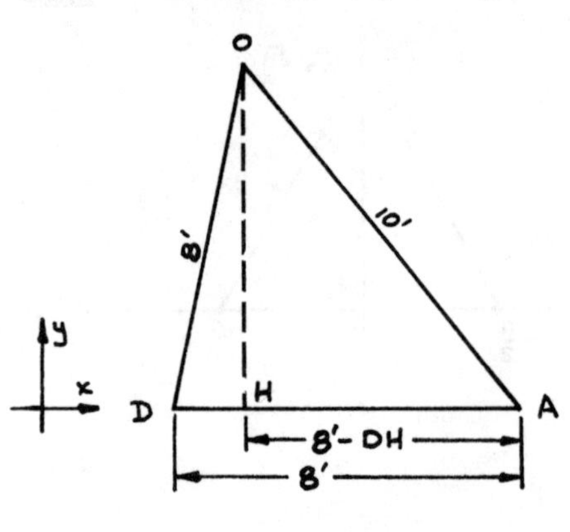

$$\overline{OH}^2 + \overline{DH}^2 = 8^2 \quad (1)$$

$$\overline{OH}^2 + (8 - DH)^2 = 10^2 \quad (2)$$

$$-\overline{OH}^2 - \overline{DH}^2 = -8^2 \quad -(1)$$

$$(8 - DH)^2 - \overline{DH}^2 = 36$$

$$64 - 16\ DH + \overline{DH}^2 - \overline{DH}^2 = 36$$

$$DH = \frac{64 - 36}{16} = 1.75 \text{ ft.}$$

$$HA = 8 - 1.75 = 6.25 \text{ ft.}$$

Solving for the distance OH

$$\overline{OH}^2 + \overline{DH}^2 = \overline{OD}^2$$

$$\overline{OH}^2 + 1.75^2 = 8^2$$

$$\overline{OH}^2 = 64 - 3.06 = 60.94$$

$$OH = 7.8 \text{ ft.}$$

With the dimensions calculated we can now solve triangle ODA, assuming that members OB and OC have been replaced by the single member OD.

$$\Sigma F_x = 0$$

$$OD_x - OA_x = 0 \qquad \frac{1.75}{8}\ OD = \frac{6.25}{10}\ OA \qquad OA = \frac{10 \times 1.75}{8 \times 6.25}\ OD$$

$$OA = 0.35\ OD$$

$$\Sigma F_y = 0$$

$$OD_y + OA_y - 1000 = 0 \qquad \frac{7.8}{8}\ OD + \frac{7.8}{10}\ OA - 1000 = 0$$

$$0.976\ OD + 0.78(0.35\ OD) = 1000$$

$$0.976\ OD + 0.273\ OD = 1000$$

$$OD = \frac{1000}{1.249} = 800 \text{ lbs.}$$

$$OA = 0.35(800) = 280 \text{ lbs.}$$

Now we can replace member OD by members OB and OC

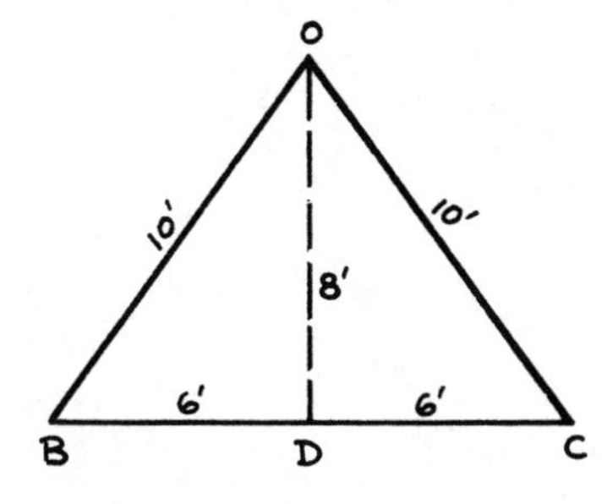

$$\Sigma F_y = 0$$

$$OB_y + OC_y = OD$$

Since OB = OC and OD = 800 lbs.,

$$2\ OB_y = 800\ ; \quad 2\left(\frac{8}{10}\right) OB = 800$$

$$OB = OC = \frac{800 \times 10}{2 \times 8} = 500 \text{ lbs.}$$

(All legs are in compression.)

STATICS 6

Find the moment of inertia of the triangle shown below about its base.

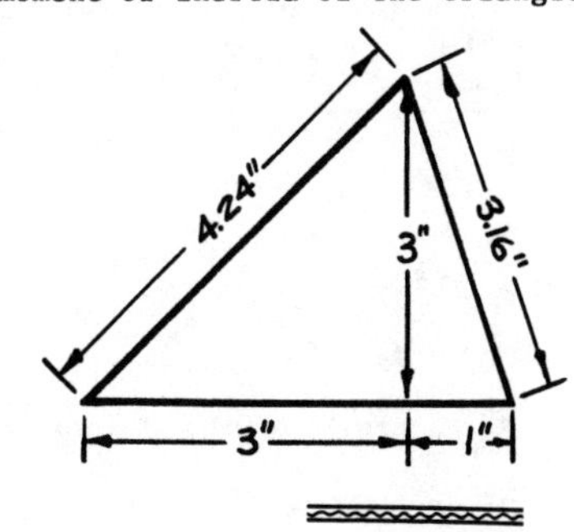

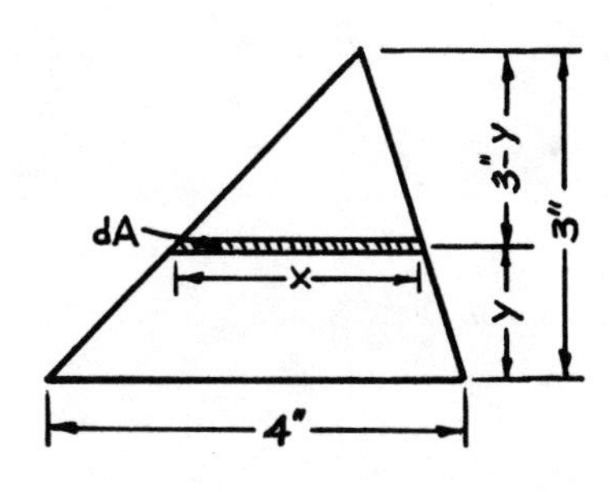

By similar triangles $\frac{x}{4} = \frac{3 - y}{3}$

$$x = \frac{4(3 - y)}{3} \qquad dA = xdy = \frac{4}{3}(3 - y)dy$$

$$I = \int y^2 dA = \frac{4}{3}\int_0^3 (3 - y)y^2 dy$$

$$= \frac{4}{3}\left[y^3 - \frac{y^4}{4}\right]_0^3 = \frac{4}{3}\left[27 - \frac{81}{4}\right]$$

$$= \frac{4}{3}(6.75) = 9 \text{ in}^4$$

STATICS 7

A weight W is suspended from the end of a 3-foot long weightless rod, as shown. Find an expression for the cosine of the angle θ when the system is in equilibrium. Neglect any friction between the contacting surfaces.

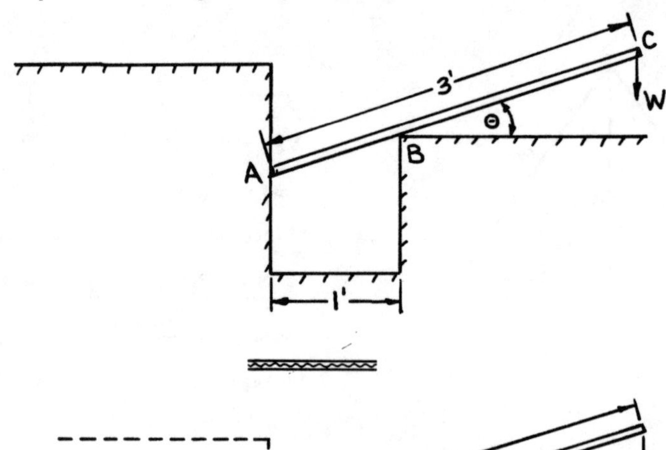

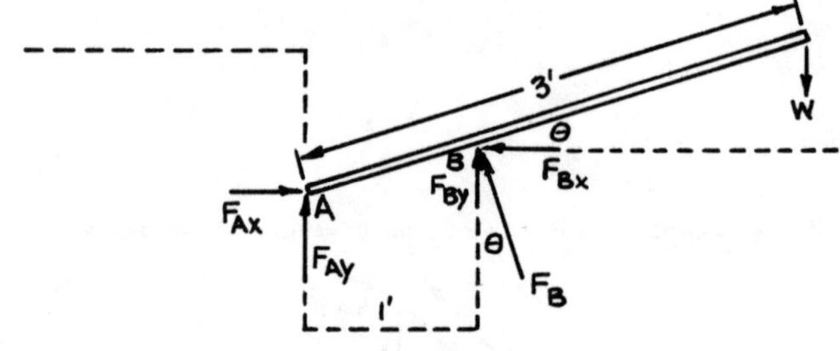

Since there is no friction $F_{A_y} = 0$

$\Sigma F_y = 0 \qquad F_{B_y} - W = 0 \qquad \therefore F_{B_y} = W$

$\Sigma F_x = 0 \qquad F_{A_x} - F_{B_x} = 0 \qquad \therefore F_{A_x} = F_{B_x}$

$\Sigma M_A = 0 \qquad (1 \times F_{B_y}) + (1 \times F_{B_x} \times \tan\theta) - (W \times 3 \times \cos\theta) = 0$

$$W + F_{B_x}\tan\theta - 3W\cos\theta = 0 \qquad (1)$$

$F_{B_x} = F_B \sin\theta \qquad F_{B_y} = F_B\cos\theta$

$$\frac{F_{B_x}}{\sin\theta} = \frac{F_{B_y}}{\cos\theta} \qquad \therefore F_{B_x} = F_{B_y}\frac{\sin\theta}{\cos\theta} = W\frac{\sin\theta}{\cos\theta}$$

Substituting the value of F_{B_x} into (1)

$$W + W\frac{\sin\theta}{\cos\theta}\frac{\sin\theta}{\cos\theta} - 3W\cos\theta = 0$$

$$W\cos^2\theta + W\sin^2\theta - 3W\cos^3\theta = 0$$

But $\cos^2\theta + \sin^2\theta = 1$

$$W - 3W\cos^3\theta = 0 \qquad \cos^3\theta = \frac{1}{3} \qquad \cos\theta = \sqrt[3]{\frac{1}{3}}$$ ●

STATICS 8

The mass moment of inertia of a cylinder about its central axis perpendicular to a circular cross section is

(a) directly proportional to its radius
(b) independent of its radius
(c) directly proportional to its length
(d) independent of its length
(e) inversely proportional to the square of its radius

mass moment of inertia $= \int r^2 dM$

for a cylinder $dM = 2\pi r dr \cdot \rho L$

where r = radial distance to dr
ρ = density of cylinder
L = length of cylinder

$$I = \int_0^R r^2 2\pi r dr \cdot \rho L = 2\pi\rho L \int_0^R r^3 dr = 2\pi\rho L \left.\frac{r^4}{4}\right|_0^R = \frac{1}{2}\pi\rho L R^4$$

Thus the mass moment of inertia is proportional to the length of the cylinder.

Answer is (c)

STATICS 9

Which of the following is true with respect to the vector diagram shown below?

(a) $\vec{D} = \vec{A} + \vec{B} - \vec{C}$

(b) $\vec{2B} = \vec{A} + \vec{2C}$

(c) $\vec{2A} = \vec{B} + \vec{D}$

(d) $\vec{C} = \vec{D} - \vec{A} - \vec{B}$

(e) $\vec{A} - \vec{B} = \vec{D} - \vec{C}$

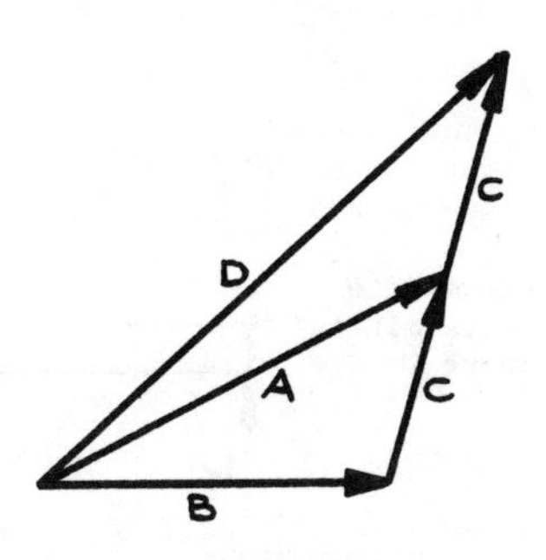

$$\vec{A} = \vec{D} - \vec{C}$$
$$\vec{A} = \vec{B} + \vec{C}$$
$$\overline{2\vec{A} = \vec{D} + \vec{B}}$$

Answer is (c)

STATICS 10

The reason at point "A" is

(a) zero
(b) 40 lbs ↑
(c) 40 lbs ↓
(d) 40 lbs ↑ plus 400 ft-lbs
(e) 40 lbs ↓ plus 400 ft-lbs

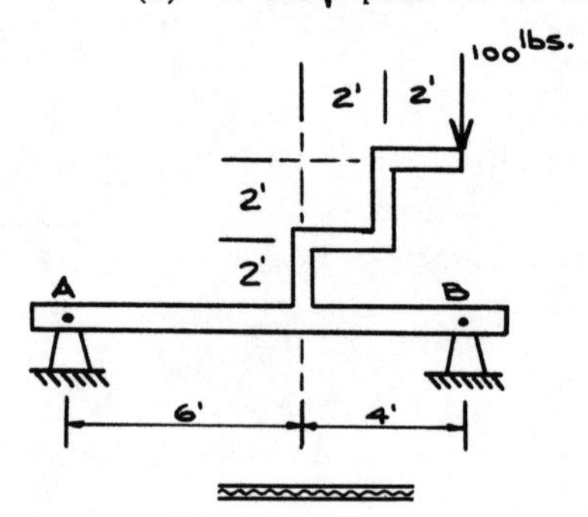

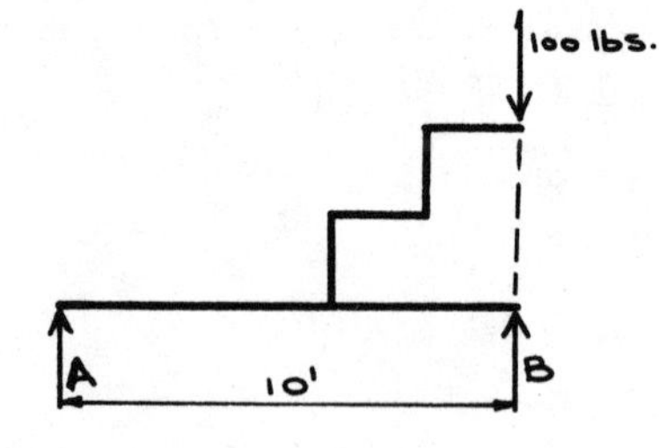

$$\Sigma M_B = 0$$
$$10A + 100(0) = 0$$
$$\therefore A = 0$$

Answer is (a)

STATICS 11

The beam shown at the right is

(a) stable
(b) unstable
(c) statically determinate
(d) statically indeterminate
(e) none of the above choices

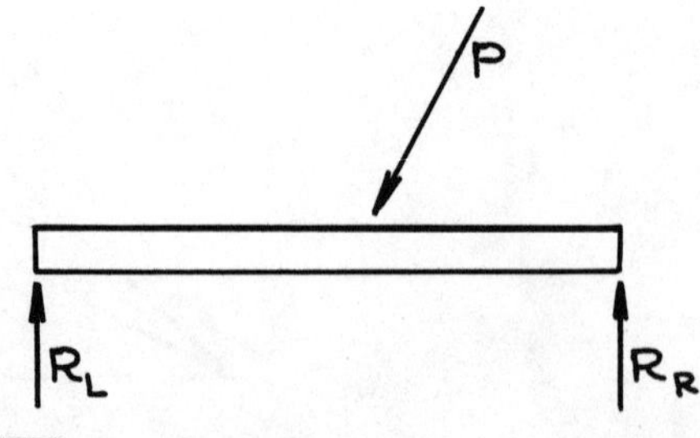

The load P has a horizontal component that is not resisted by the reactions R_L or R_R. Therefore the beam will slide and is thus unstable.

Answer is (b)

STATICS 12

An airtight closed box of weight P is suspended from a spring balance. A bird of weight W is placed on the floor of the box, and the balance reads W + P. If the bird flies around in the box at a constant elevation without accelerating, what is the balance reading?

(a) P
(b) P - W
(c) P + 2W
(d) P + W/2
(e) P + W

$$F = P + \frac{W}{2} + \frac{W}{2} = P + W$$

Answer is (e)

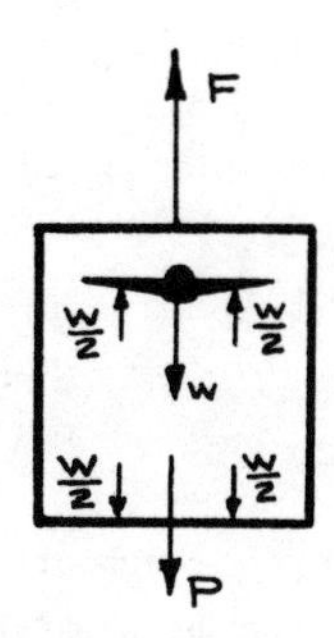

STATICS 13

Given the truss as shown below. Determine mathematically the stress in members U_2U_3, U_2L_3, L_2L_3, and indicate if the member is in tension or compression.

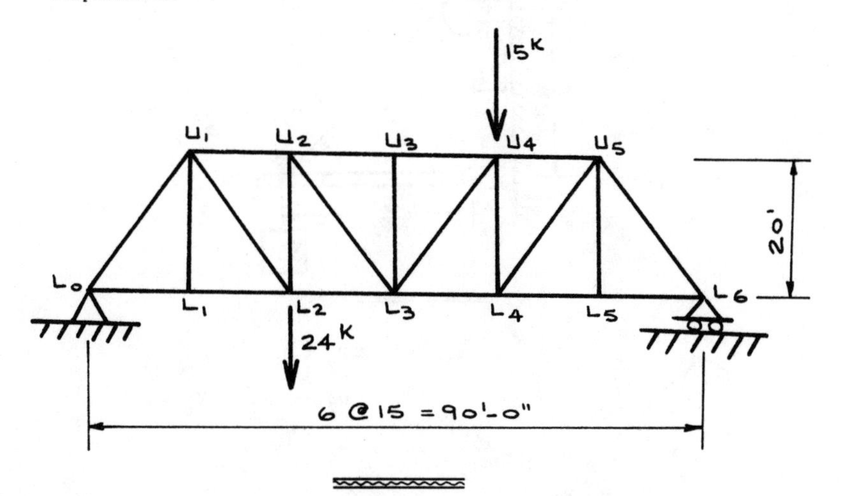

Compute R_{L_0} and R_{L6}

$\Sigma M_{L_0} = 0 \qquad (6 \times 15)R_{L6} - 15(4 \times 15) - 24(2 \times 15) = 0$

$R_{L6} = \frac{900 + 720}{90} = 18^k \qquad R_{L_0} = 24 + 15 - 18 = 21^k$

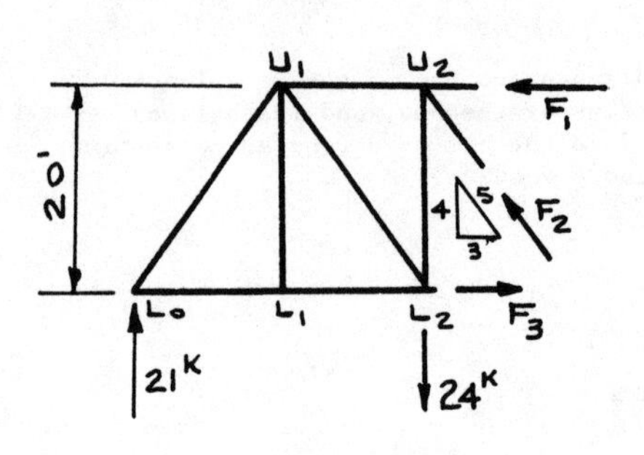

$\Sigma F_y = 0$

$$F_2\left(\tfrac{4}{5}\right) + 21 - 24 = 0$$

$$F_2 = \frac{5(3)}{4} = 3.75^k$$

● $U_2L_3 = 3.75^k$ Comp.

$\Sigma M_{U2} = 0$

$$20F_3 - 21(30) = 0$$

$$F_3 = 31.5^k$$

● $L_2L_3 = 31.5^k$ Tens.

$\Sigma F_x = 0$

$$F_1 + F_2\left(\tfrac{3}{5}\right) - F_3 = 0$$

$$F_1 = 31.5 - 3.75\left(\tfrac{3}{5}\right)$$

$$F_1 = 31.5 - 2.25 = 29.25^k$$

● $U_2U_3 = 29.25^k$ Comp.

STATICS 14

A weight W_1 is supported by a rigid, weightless bar, AB, and a weightless cable loaded by a weight W_2, as shown. Find the relationship between W_1 and W_2 for equilibrium of the system, with the bar AB in a horizontal position, and show that this position of equilibrium is unstable. Neglect friction, and assume negligible dimensions for the pulley.

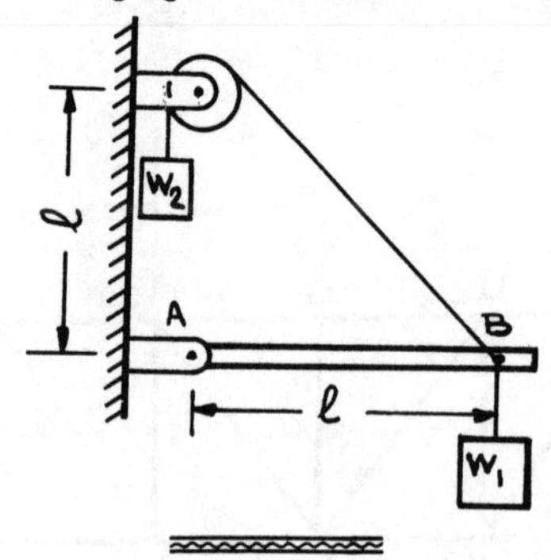

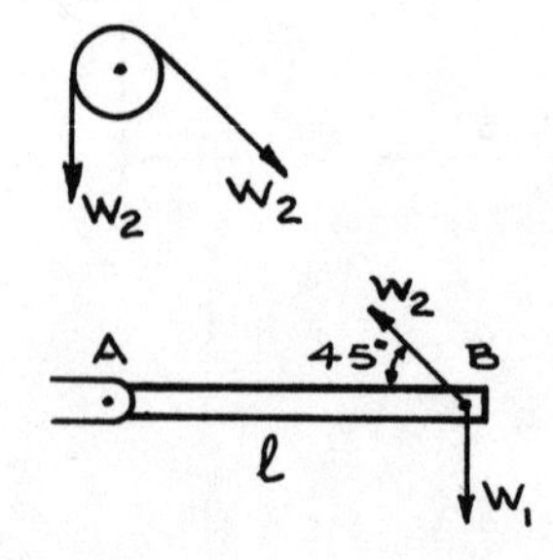

$\Sigma F_y = 0$

$$W_2 \cos 45° - W_1 = 0$$

$$W_1 = \frac{\sqrt{2}}{2} W_2$$

or $W_2 = \sqrt{2}\, W_1$ ●

In stable equilibrium small movements of the structure produce forces that tend to restore the structure to its original condition.

In unstable equilibrium small movements produce forces that tend to cause still further movement.

In this problem assume the structure is deflected by displacing B until bar AB makes a small angle ϕ with the horizontal. We can then compute the sum of the moments about joint A to see what the effect is on the structure.

If B deflected downward	If B deflected upward
clockwise moment = $W_1L\cos\phi$	clockwise moment = $W_1L\cos\phi$
counterclockwise moment: $\sqrt{2}\,W_1L\sin(45 - \frac{\phi}{2})$	counterclockwise moment: $\sqrt{2}\,W_1L\sin(45 + \frac{\phi}{2})$

if ϕ small, say 5°

$\cos\phi = 0.996$

$\sin(45 - \frac{\phi}{2}) = 0.676$

$\sin(45 + \frac{\phi}{2}) = 0.737$

$0.996W_1L > 0.956W_1L$	$0.996W_1L < 1.04W_1L$
CW moment > CCW moment	CW moment < CCW moment
Unstable	Unstable

The structure is thus in unstable equilibrium. ●

STATICS 15

A piece of metal plate in the shape of a right triangle is suspended as shown below. If its weight is W, and its sides are a and b, what is the force at string A?

(a) W/3
(b) W/2
(c) 2W/3
(d) 3W/4
(e) W

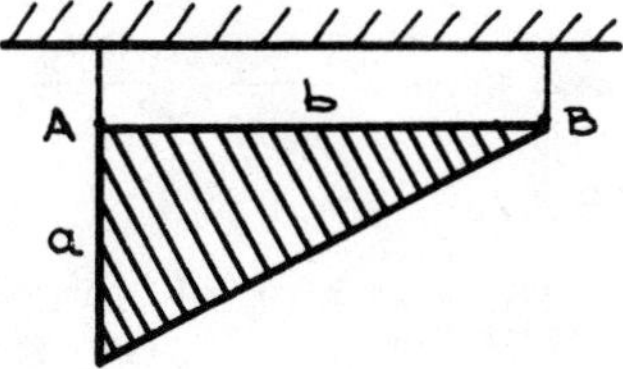

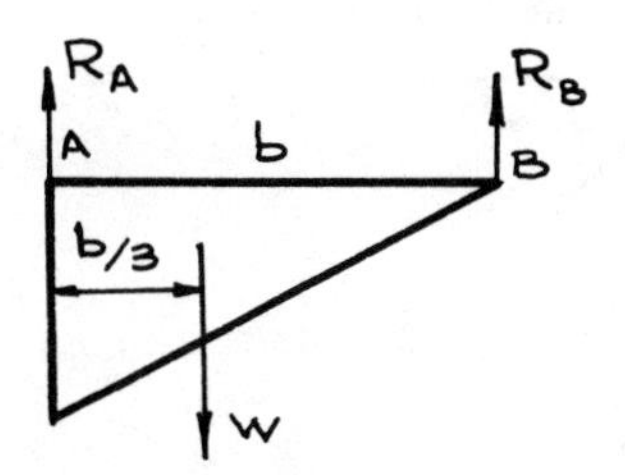

$\Sigma M_B = 0 \qquad R_A(b) - W(\frac{2b}{3}) = 0$

$R_A = \frac{2bW}{3b} = \frac{2W}{3}$ ●

Answer is (c)

STATICS 16

Given a triangle as shown. Find the percentage error in the X coordinate of the centroid if five equal width rectangles are used to approximate the area. Assume line cd goes through the mid-point of each rectangle.

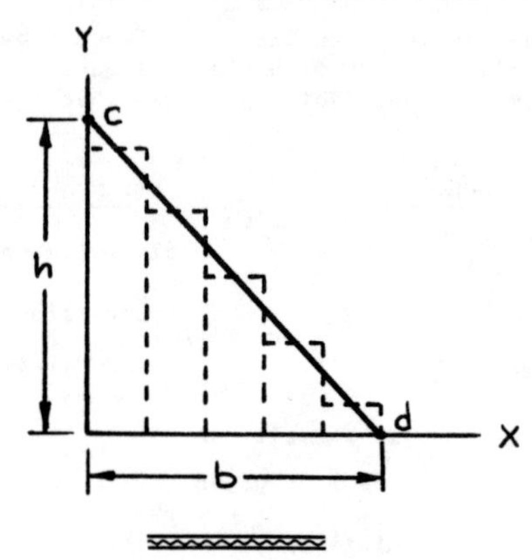

Compute the centroid of the 5 rectangles:

Rectangle	Height	Width	Area	Distance to Centroid	Ax
1	0.9h	0.2b	0.18bh	0.1b	$0.018b^2h$
2	0.7h	0.2b	0.14bh	0.3b	$0.042b^2h$
3	0.5h	0.2b	0.10bh	0.5b	$0.050b^2h$
4	0.3h	0.2b	0.06bh	0.7b	$0.042b^2h$
5	0.1h	0.2b	0.02bh	0.9b	$0.018b^2h$
			0.50bh		$0.170b^2h$

$$\bar{x} = \frac{\Sigma Ax}{\Sigma A} = \frac{0.170b^2h}{0.50bh} = 0.34b$$

We know, however, that the centroid of a triangle is located at $\frac{1}{3}b$

$$\text{Percent Error} = \frac{0.34b - 0.333b}{0.333b} = \frac{0.007b}{0.333b} = 2.1 \text{ percent}$$ ●

STATICS 17

The reaction at B to hold the weightless hinged frame on the next page in equilibrium is:

(a) 333 lbs →

(b) 333 lbs → plus 800 in. lbs. ↻

(c) 333 lbs → plus 1200 in. lbs. ↻

(d) 333 lbs ↘ 36.9°

(e) 333 lbs

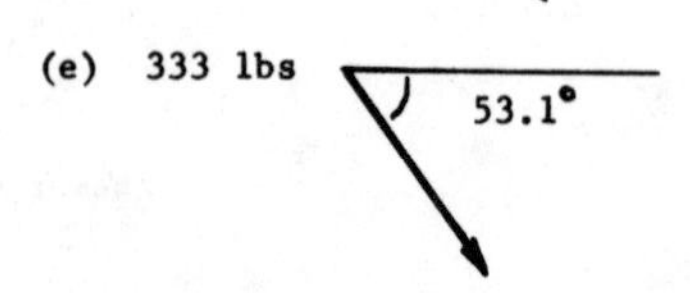

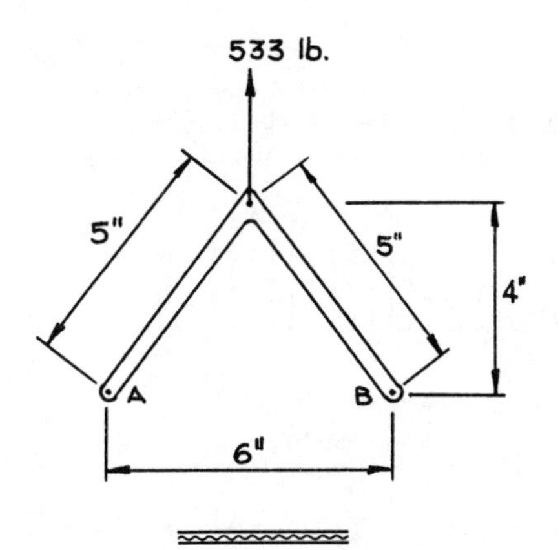

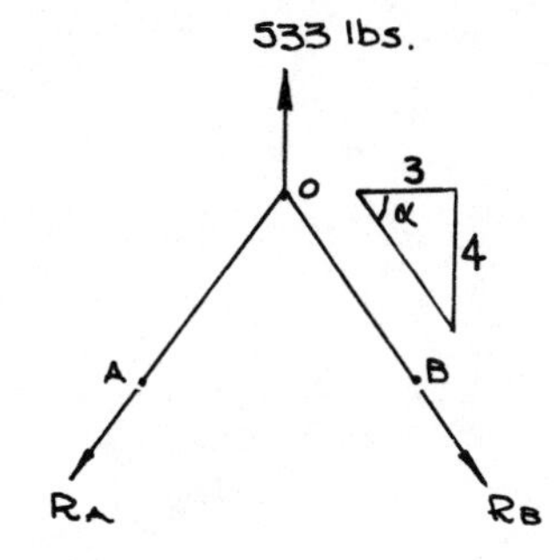

The line of action of all forces must pass through point O. Therefore the direction of R_B is

$\alpha = \tan^{-1}\frac{4}{3} = 53.1^\circ$ with respect to the horizontal.

There must be a similar reaction at A.

The magnitude of reactions A and B:

$R_A = R_B = \frac{533}{2} \times \frac{5}{4} = 333$ lbs.

Answer is (e)

STATICS 18

The moment of inertia of any plane figure can be expressed in units of length to the

(a) First power
(b) Second power
(c) Third power
(d) Fourth power
(e) Fifth power

For a rectangular cross-section about its centroidal axis

$$I_x = \int y^2\, dA = \int_{-h/2}^{+h/2} y^2 b\, dy = b\left[\frac{y^3}{3}\right]_{-h/2}^{+h/2} = \frac{bh^3}{12}$$

Thus the moment of inertia is a fourth power of length.

Answer is (d)

STATICS 19

For the beam loaded as shown, express the reactions at A and B by algebraic expressions in terms of K and L.

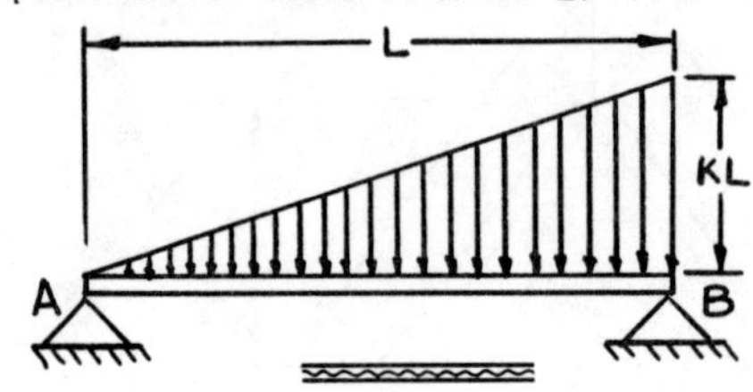

$\Sigma M_A = 0 \qquad LB - \frac{1}{2}L(KL)(\frac{2}{3}L) = 0 \qquad B = \frac{KL^2}{3}$ ●

$\Sigma M_B = 0 \qquad -LA + \frac{1}{2}L(KL)(\frac{1}{3}L) = 0 \qquad A = \frac{KL^2}{6}$ ●

As a check

$\Sigma F_y = 0 \quad A + B - \frac{1}{2}KL = 0 \quad \frac{1}{6}KL + \frac{1}{3}KL - \frac{1}{2}KL = 0 \quad 0 = 0$

STATICS 20

The vector which represents the sum of a group of force vectors is called the

(a) magnitude
(b) resultant
(c) sum
(d) phase angle
(e) force polygon

A vector sum is called the resultant vector. ●

Answer is (b)

STATICS 21

Which of the following is not a vector quantity?

(a) velocity
(b) speed
(c) acceleration
(d) displacement
(e) momentum

Speed is a rate $\frac{ds}{dt}$ and has a magnitude but not direction. It is not a vector. ●

Answer is (b)

STATICS 22

A horizontal equilateral plate ABC (shown below), 10 feet on a side, is supported at the vertices. What are the reactions at the supports due to a load of 500 pounds acting at a point on the median line 3 feet from vertex A?

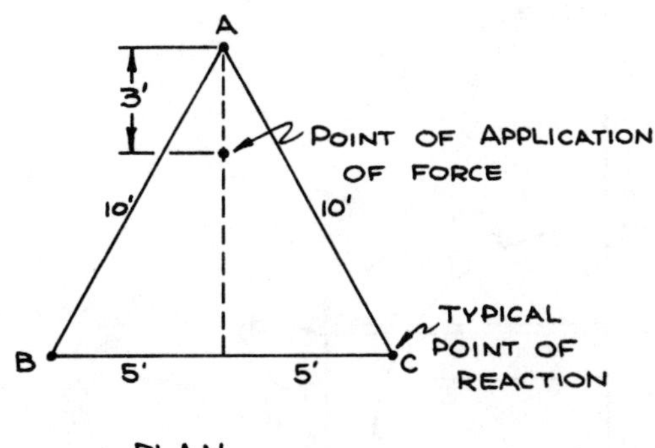

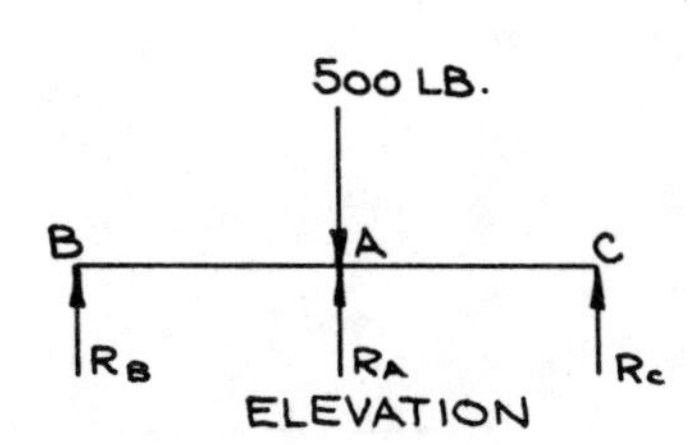

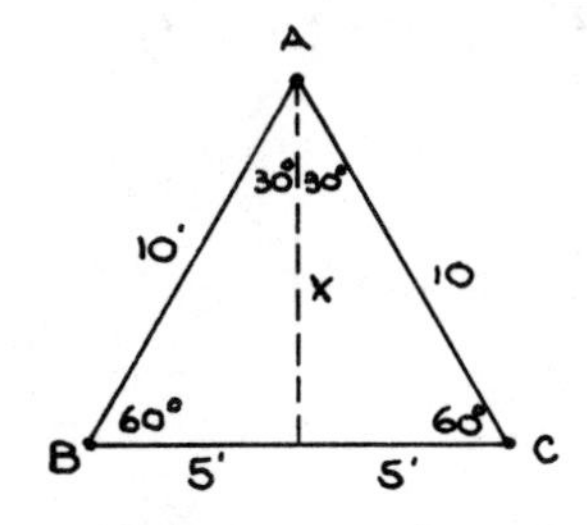

X = 8.66'
5.66'
500 lb.
3'
B,C
$R_{B,C}$
R_A

$\cos 30^\circ = \frac{X}{10} = 0.866$

Therefore $X = 8.66$ ft

$\Sigma M_A = 0$

$R_{B,C}(8.66) - 500(3) = 0$ $\qquad R_{B,C} = \frac{3}{8.66}(500) = 173$ lbs

$R_B = R_C = \frac{173}{2} = 86.5$ lbs ●

$\Sigma F_y = 0$

$R_{B,C} + R_A - 500 = 0$ $\qquad 173 + R_A - 500 = 0$

$R_A = 327$ lbs ●

STATICS 23

Find the centroid of a solid hemisphere of radius "r".

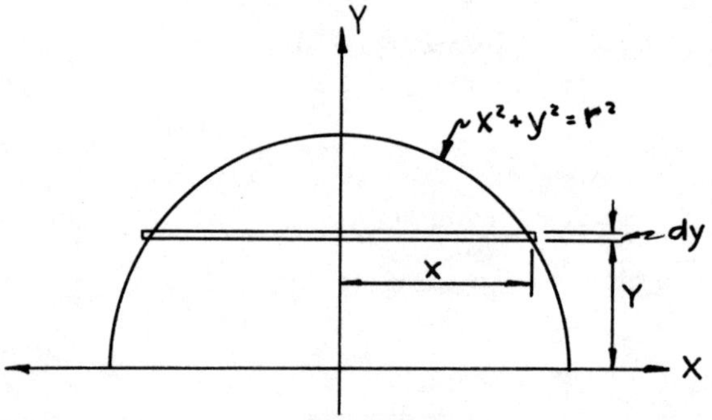

$$\bar{Y} = \frac{\int Y\,dV}{\int dV} \qquad X^2 + Y^2 = r^2 \quad \text{or} \quad X^2 = (r^2 - Y^2)$$

$$dV = \pi(r^2 - Y^2)dy$$

$$\int Y\,dV = \int_0^r Y\pi(r^2 - Y^2)dy = \pi\int_0^r (Yr^2 - Y^3)dy = \pi\left|\frac{Y^2r^2}{2} - \frac{Y^4}{4}\right|_0^r$$

$$= \pi\left(\frac{r^4}{2} - \frac{r^4}{4}\right) = \pi\frac{r^4}{4}$$

$$\int dV = \pi\int_0^r (r^2 - Y^2)dy = \pi\left|r^2Y - \frac{Y^3}{3}\right|_0^r = \pi(r^3 - \frac{r^3}{3}) = \frac{2}{3}\pi r^3$$

$$\bar{Y} = \frac{\int Y\,dV}{\int dV} = \frac{\pi\frac{r^4}{4}}{\frac{2}{3}\pi r^3} = \frac{3}{8}r$$

Location of Centroid: $\bar{Y} = \frac{3}{8}r$ ●

By symmetry $\bar{X} = 0$ ●

3

Dynamics

DYNAMICS 1

If the disk shown below is rolling without slipping, what is the direction of the velocity of point p relative to point q at the instant shown?

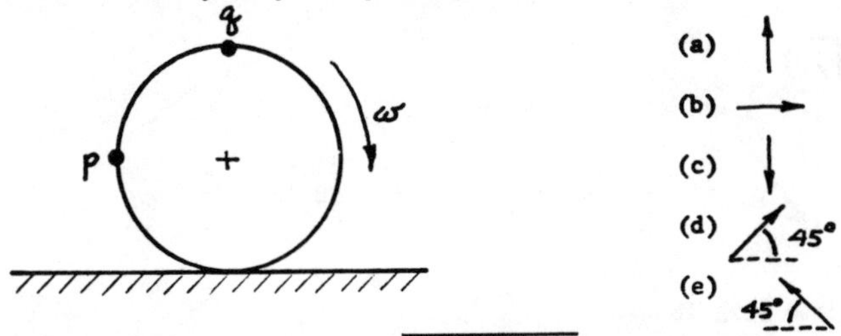

Since p and q both lie on the disk, the facts that the disk contacts a surface and that it rolls without slipping are unrelated to motion of p relative to q. p and q have equal tangential velocities, p directed upward and q directed to the right.

For motion of p relative to q, consider q stationary and add q's reversed velocity vector to p's vector. The resultant is motion of p relative to (fixed) q.

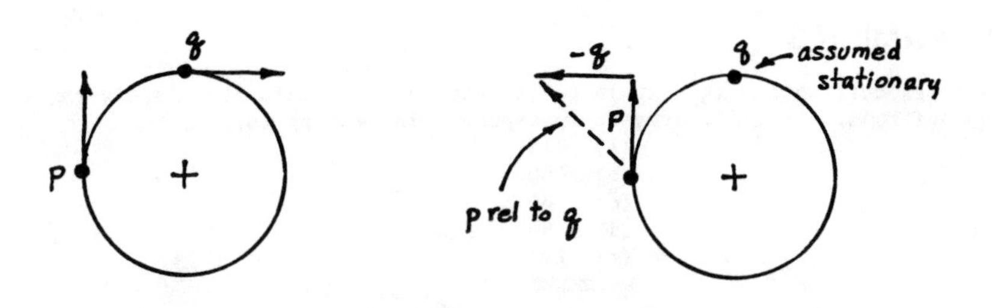

Answer is (e)

DYNAMICS 2

A 3.22-pound bead slides down a wire without friction along a circular path in a vertical plane as shown below. The speed of the bead along the wire in the position shown is 10.0 feet per second. The magnitude of the force of the wire on the bead in this position is:

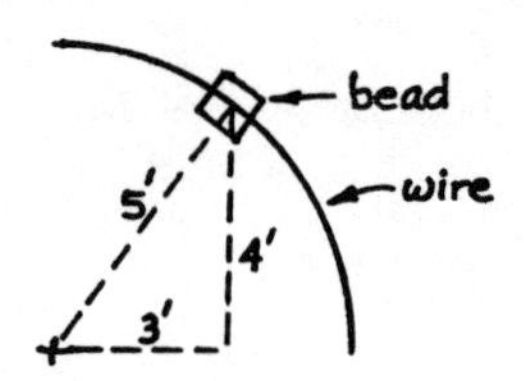

(a) 0.576 lb.
(b) 1.93 lb.
(c) 2.00 lb.
(d) 2.57 lb.
(e) 3.22 lb.

Since there is no friction between bead and wire, there is no tangential component of bead weight exerted on the wire. Total force on wire is the vector sum of the radial component of bead weight and the radial force caused by its tangential velocity of 10 ft/sec.

$$F_{radial} = \frac{mv_T^2}{r} = \frac{3.22(10)^2}{32.2(5)} = 2.00 \text{ lb} \nearrow$$

Radial and tangential components of bead weight are:

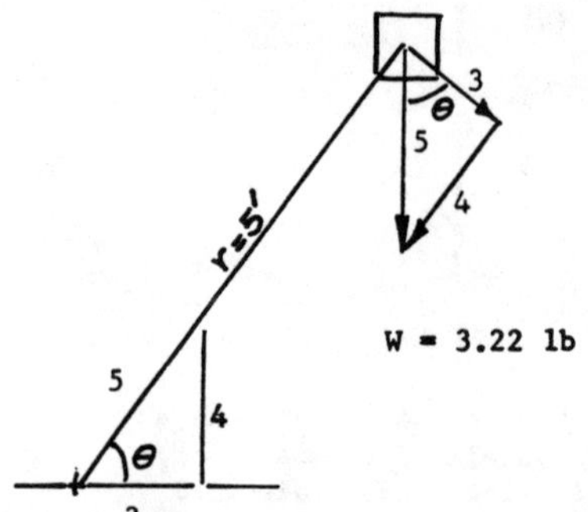

tangential $= \frac{3}{5}(3.22) = 1.93$ lb $\searrow$

radial $= \frac{4}{5}(3.22) = 2.58$ lb $\swarrow$

W = 3.22 lb

Net force of bead on wire = 2.58 lb $\swarrow$ + 2.00 lb $\nearrow$ = 0.58 lb $\swarrow$

Reaction force of wire on bead = 0.58 lb $\nearrow$ ●

Answer is (a)

DYNAMICS 3

A large coil spring (k = 10 lb/in) is elongated, within its elastic range, by one foot. Stored energy in foot-pounds is nearest to:

(a) 10
(b) 40
(c) 60
(d) 80
(e) 120

F = kx, k = 10 lb/in = 120 lb/ft.

$$\text{Energy, } E = \int_0^1 F\,dx = \int_0^1 kx\,dx = 120 \left.\frac{x^2}{2}\right]_0^1 = 60 \text{ ft lb}$$ ●

Answer is (c)

DYNAMICS 4

A ball thrown from A with velocity 17 ft/sec strikes a fixed barrier at B, and rebounds to C at a velocity of 16.25 ft/sec. ϕ is unequal to ϕ'

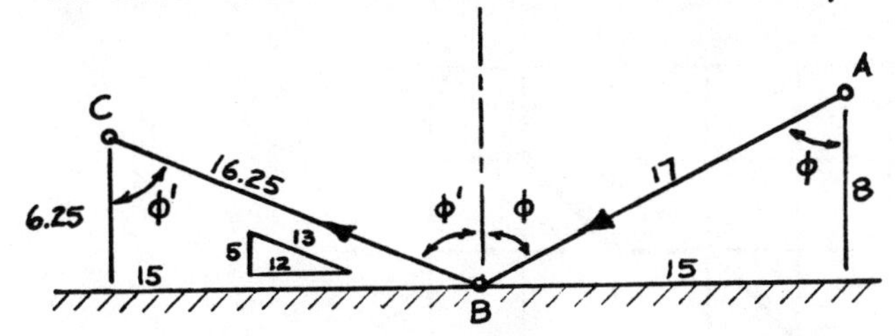

The coefficient of restitution is nearest to:

(a) 0.74
(b) 0.78
(c) 0.83
(d) 0.92
(e) 0.96

In this particular situation, coefficient of restitution $e = \dfrac{\tan\ \phi}{\tan\ \phi'}$

where ϕ is the incident angle and ϕ' is the emergent angle.

$$e = \frac{15/8}{12/5} = \frac{25}{32} = 0.78$$

In the usual case where both impacting objects move:

$$e = \frac{\text{relative velocity after impact}}{\text{relative velocity before impact}} = \frac{v_2' - v_1'}{v_1 - v_2}$$

For oblique impact the velocities used must be the components of velocity in a direction normal to the point of contact, in this case, the y direction.

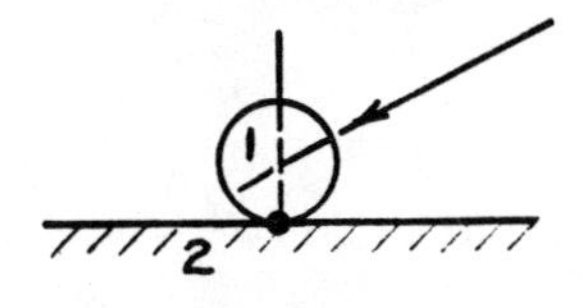

$$e = \frac{0 - (+6.25)}{-8 - 0} = 0.78$$ ●

Answer is (b)

DYNAMICS 5

The cords and frictionless pulley, of the system shown in the drawing, have negligible mass. The distance in feet that the 32.2 lb mass will move during the first half second after release from rest is nearest to:

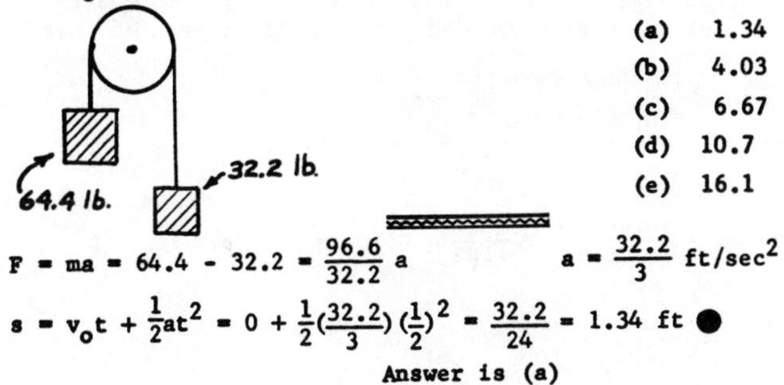

(a) 1.34
(b) 4.03
(c) 6.67
(d) 10.7
(e) 16.1

$$F = ma = 64.4 - 32.2 = \frac{96.6}{32.2}a \qquad a = \frac{32.2}{3}\ \text{ft/sec}^2$$

$$s = v_o t + \frac{1}{2}at^2 = 0 + \frac{1}{2}\left(\frac{32.2}{3}\right)\left(\frac{1}{2}\right)^2 = \frac{32.2}{24} = 1.34\ \text{ft}$$ ●

Answer is (a)

DYNAMICS 6

Given the velocity-time plot for motion shown below:

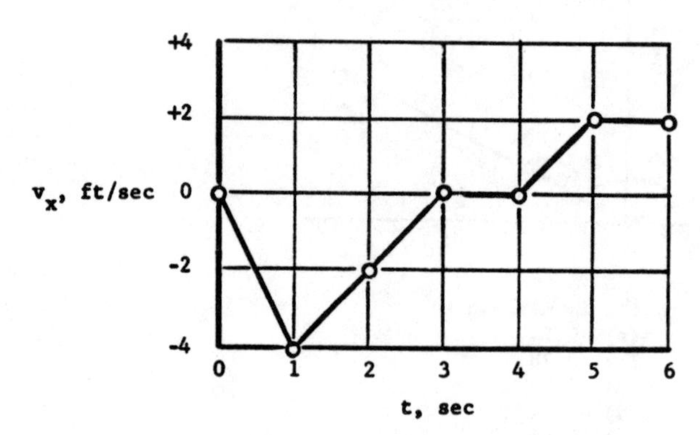

Displacement of the object from t = 1 sec to t = 6 sec will be:

(a) 4 feet to left
(b) 1 foot to left
(c) zero
(d) 1 foot to right
(e) 3 feet to right

$S_x = \int_1^6 v_x dt.$ Graphically integrate by summing positive and negative areas, above and below the zero velocity line, between t = 1 and t = 6.

$S_x = -4 + 1 + 2 = -1$ (to left) ●

Total travel will have been 7 feet (4 to left and 3 to right).

Answer is (b)

DYNAMICS 7

A hoist with a 100-horsepower engine is capable of lifting a 10,000 lb load a height of 20 feet in 10 seconds. What is the efficiency of this machine?

$$\text{output horsepower} = \frac{10{,}000 \text{ lbs} \times \frac{20 \text{ feet}}{10 \text{ sec.}}}{550 \ \frac{\text{ft-lb}}{\text{sec.}}} = 36.4 \text{ hp}$$

$$\text{efficiency} = \frac{\text{output hp}}{\text{input hp}} = \frac{36.4}{100} = 0.364 = 36.4 \text{ per cent}$$ ●

DYNAMICS 8

An object is placed 3 feet from the center of a horizontally rotating platform. The coefficient of friction is 0.30. The object will begin to slide off when the platform speed is nearest to:

(a) 12 rpm
(b) 17 rpm
(c) 22 rpm
(d) 26 rpm
(e) 33 rpm

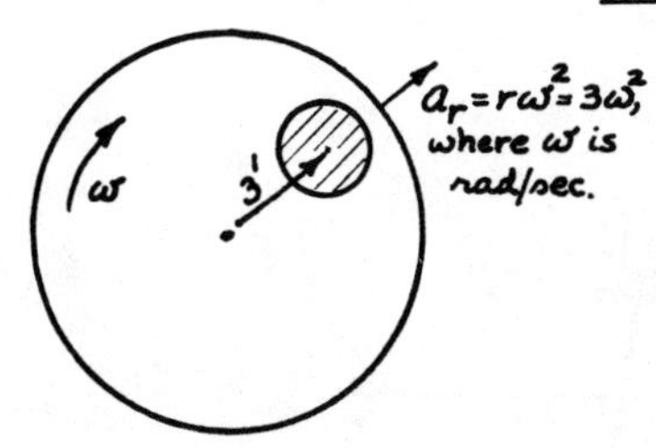

W = weight of object

$$F_r - F_F = 0 = \frac{W}{32.2}\,3\omega^2 - 0.3W$$

$$0 = \frac{3}{32.2}\omega^2 - 0.3$$

$$\frac{3\omega^2}{32.2} = 0.3 \qquad \therefore\ \omega^2 = \frac{0.3(32.2)}{3}$$

$$\omega = \sqrt{3.22} = 1.79 \text{ rad/sec}$$

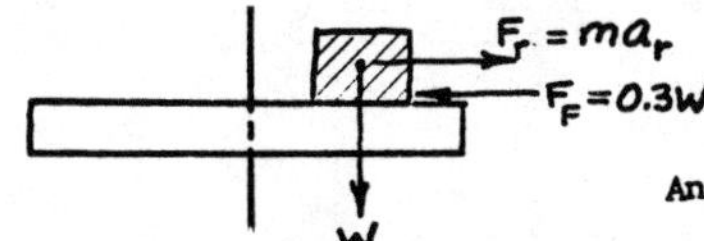

$$1.79\ \frac{\text{rad}}{\text{sec}} \times \frac{60 \text{ sec}}{\text{min}} \times \frac{\text{rev}}{2\pi \text{ rad}} = 17.1\ \frac{\text{rev}}{\text{min}}$$ ●

Answer is (b)

DYNAMICS 9

Two barges, one weighing 10 tons, the other weighing 20 tons are connected by a cable in quiet water. Initially the barges are at rest 100 feet apart. The cable is reeled in until the barges are 50 feet apart. If the friction is negligible, calculate the distance moved by the 10 ton barge.

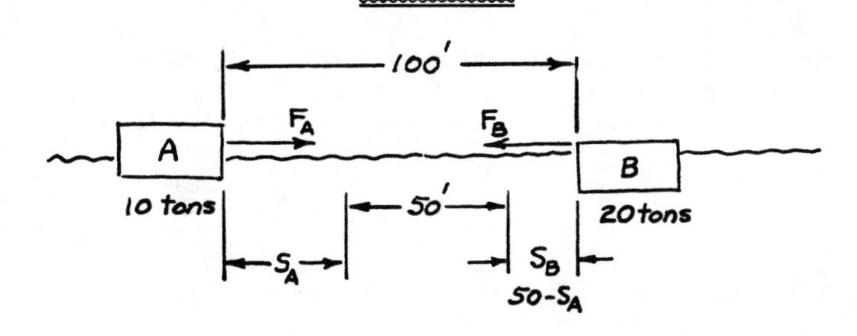

Equal and opposite forces are exerted on A and B when the cable is reeled in. Assuming constant force, accelerations are constant, but unequal.

$$F_A = F_B = m_A a_A = m_B a_B \qquad \frac{20{,}000}{32.2}\,a_A = \frac{40{,}000}{32.2}\,a_B \qquad \therefore\ a_A = 2a_B$$

At any time t, the distances traveled from rest are:

$$s_A = \frac{1}{2}\,a_A t^2 \quad \text{and} \quad s_B = \frac{1}{2}\,a_B t^2$$

Therefore

$$\frac{s_A}{s_B} = \frac{a_A}{a_B} = \frac{2a_B}{a_B} = \frac{2}{1} \qquad \text{Since } s_A + s_B = 50 \text{ ft}, \qquad s_A = \frac{2}{3}(50) = 33.3 \text{ ft}$$ ●

DYNAMICS 10

A 60-ton car is to travel around a curve of radius 3,000 ft. at a speed of 60 mph. By how much must the outer rail be elevated in order that the reaction against the track may be perpendicular to the plane of the rails? The track is standard gage of 4'-8 1/2".

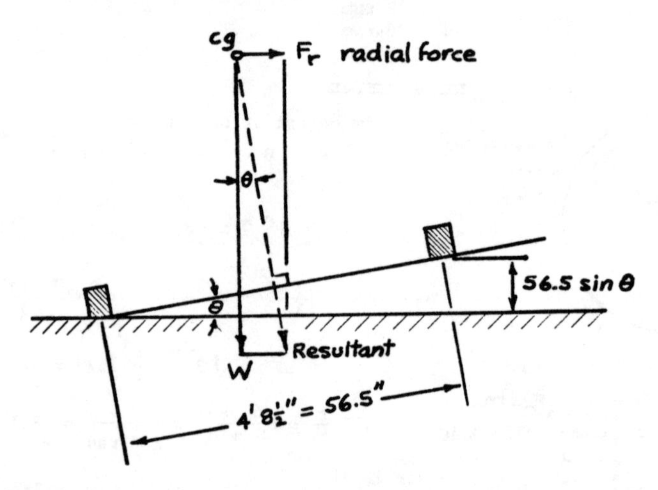

Car weight W is directed vertically downward through its cg.
Reaction against track is the resultant of radial force and car weight.
Resultant must be perpendicular to the plane of the rails.

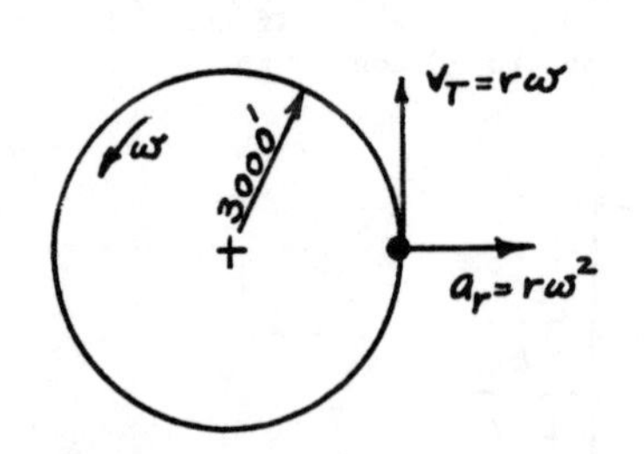

Tangential velocity, v_T, is 60 mph = 88 ft/sec.

Radial acceleration $a_r = r\omega^2$

Radial Force $F_r = ma_r = mr\omega^2 = \frac{mv_T^2}{r}$

$$F_r = \frac{W\,(88)^2}{32.2(3000)} = \frac{7{,}750W}{96{,}600}$$

$$= 0.08W$$

$\tan\theta = \frac{0.08W}{W} = 0.08$ (For small angles $\sin\theta = \tan\theta$)

Therefore,

Elevation = 56.5(0.08) = 4.53" ●

DYNAMICS 11

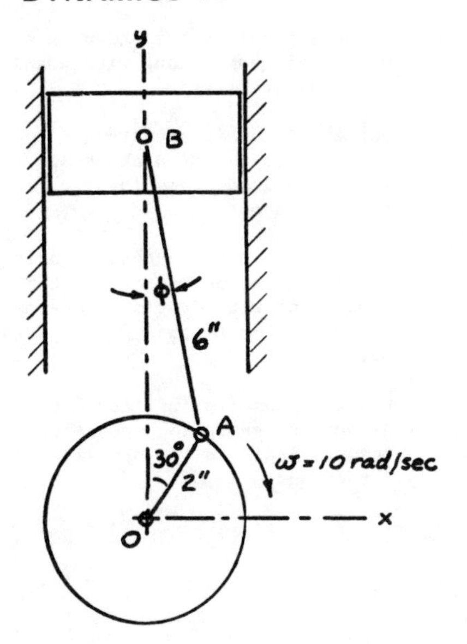

Dimensions of the reciprocating engine in the drawing are crank equals 2 inches and connecting rod equals 6 inches.

The crank rotates at a constant angular velocity of 10 rad/sec.

Find the instantaneous velocity and acceleration of the piston at the position shown.

This is a problem involving combined rotation, and rotation plus translation.

Since the journal bearing at A connects crank OA and connecting rod AB, motion of A derived from its constant velocity rotation about O is the same as motion obtained from translation of piston B plus rotation of the connecting rod AB about wrist pin B.

Solution depends upon equating x and y components that result both from translation of B and from A's rotation about B.

Angle ϕ is determined by law of sines for triangle OAB:

$\frac{\sin\phi}{2''} = \frac{\sin 30^\circ}{6''}$ $\quad \sin\phi = \frac{2}{6}(0.50) = \frac{1}{6} = 0.167$, and from a right triangle of sides 1, 6 and $\sqrt{35}$: $\quad \cos\phi = \frac{\sqrt{35}}{6} = 0.987$

Curvilinear motion involves normal N (radial) and tangential T components of velocity and acceleration:

$v_T = r\omega \qquad\qquad a_T = r\alpha$

$(\alpha = 0$ at constant $\omega)$

$v_N = \frac{dr}{dt} = 0$, at constant radius

$a_N = r\omega^2$, directed toward center of rotation

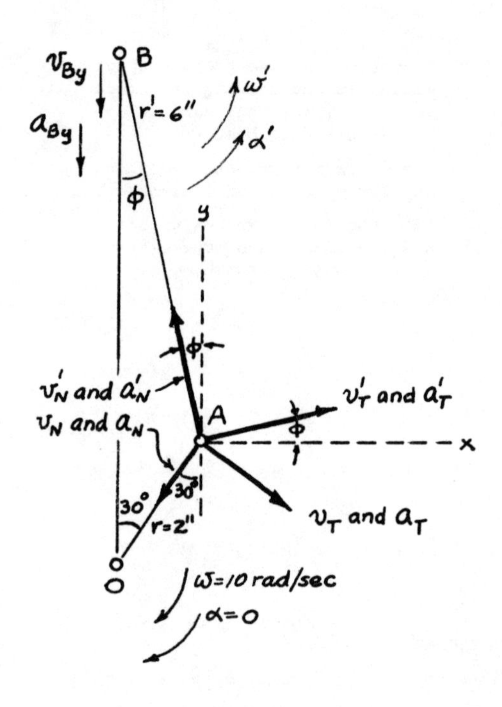

These components are shown on the vector diagram, along with x and y coordinate axes.

Primes (') designate items involved in rotation about B, while unprimed items involve rotation about O.

Directions of tangential velocity and acceleration vectors must be consistent with stated or assumed directions of α and ω producing them.

Positive answers indicate correct direction assumption; negative answers indicate the opposite direction.

The x component of velocity of A, v_{Ax}, results from sum of x components of v_T and v_N. v_{Ax} also equals sum of x components of v_T' and v_N' plus x component of B's translation, v_{Bx}. Since radii are fixed and B is constrained to only translate vertically, v_N, v_N', and v_{Bx} are zero.

$$v_{Ax} = v_T \cos 30^\circ - v_N \sin 30^\circ = v_T' \cos \phi - v_N' \sin \phi + v_{Bx}$$

$$r\,\omega \cos 30^\circ - 0 = r'\,\omega' \cos \phi - 0 + 0$$

$$2(10)(0.866) = 6(\omega')(0.987)$$

$$\omega' = +\,2.93 \text{ rad/sec} \circlearrowright \quad \text{(angular velocity at B)}$$

Similarly, y-component of A's velocity equals sum of y components of v_T and v_N, and also equals sum of y components of v_T' and v_N' plus y component of B's translation.

$$v_{Ay} - v_T \sin 30^\circ - v_N \cos 30^\circ = v_T' \sin \phi + v_N' \cos \phi - v_{By}$$

$$-\,r\,\omega \sin 30^\circ - 0 = r'\,\omega' \sin \phi + 0 - v_{By}$$

$$-2(10)(0.500) = 6(2.93)(0.167) - v_{By}$$

$$v_{By} = +\,12.93 \text{ in/sec} \downarrow \quad \bullet \text{ (piston velocity)}$$

Accelerations are similarly determined from x and y components at A. α is zero, but α' is not.

x component: $a_T \cos 30^o - a_N \sin 30^o = a_T' \cos \phi - a_N \sin \phi + a_{Bx}$

$$r\alpha \cos 30^o - r\omega^2 \sin 30^o = r'\alpha' \cos \phi - r'(\omega')^2 \sin \phi + 0$$

$$2(0)(0.866) - 2(10)^2(0.500) = 6(\alpha')(0.987) - 6(2.93)^2(0.167) + 0$$

$$\alpha' = -15.45 \text{ rad/sec}^2$$

(angular acceleration at B)

y component: $-a_T \sin 30^o - a_N \cos 30^o = a_T' \sin\phi + a_N' \cos \phi - a_{By}$

$$0 \sin 30^o - r\omega^2 \cos 30^o = r'\alpha' \sin\phi + r'(\omega')^2 \cos \phi - a_{By}$$

$$0 - 2(10)^2(0.866) = 6(-15.45)(0.167) + 6(2.93)^2(0.987) - a_{By}$$

$$a_{By} = +208.5 \text{ in/sec}^2 \downarrow$$ ●

(piston acceleration)

DYNAMICS 12

A system of four particles of equal mass (m) rotate with an angular velocity (ω) at equal distances (L) from the center of rotation, and they are spaced at equal angles as shown.
Find the magnitude and direction of the vector, $\vec{H}_c$, representing the moment of momentum of the system about the point of rotation.

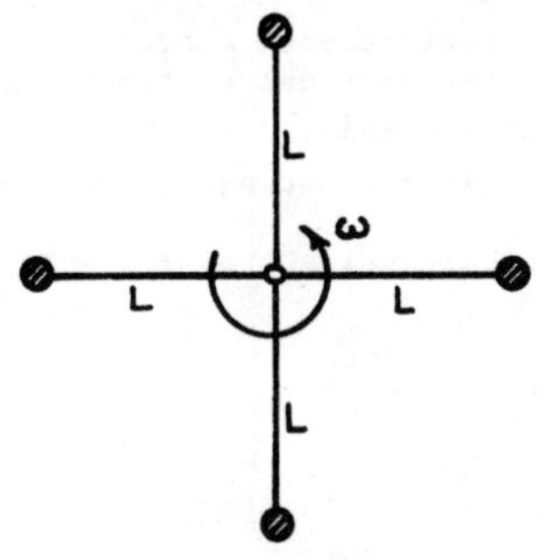

$\vec{H}_c$ = constant vector

$$= 4(\vec{\rho} \times m\vec{V}) = 4m\rho V \sin 90^\circ = 4m\rho(\rho\omega)$$

Since $\rho = L$ $\therefore \vec{H}_c = 4mL^2\omega$ ●

$\vec{H}_c$ is perpendicular to the plane of ρ and V, and is directed outward from the figure, i.e., toward the reader. ●

DYNAMICS 13

A 10-ton space craft orbits the earth at a mean altitude of 200 miles with a mean tangential velocity of 17,000 mph. At what velocity will the craft orbit the moon at the same elevation above the surface?

Diameter of the earth - - - - 8,000 miles
Diameter of the moon - - - - 2,000 miles
Mass of the earth - - - - - - 5.4×10^{21} tons
Mass of the moon - - - - - - 1/81 that of the earth

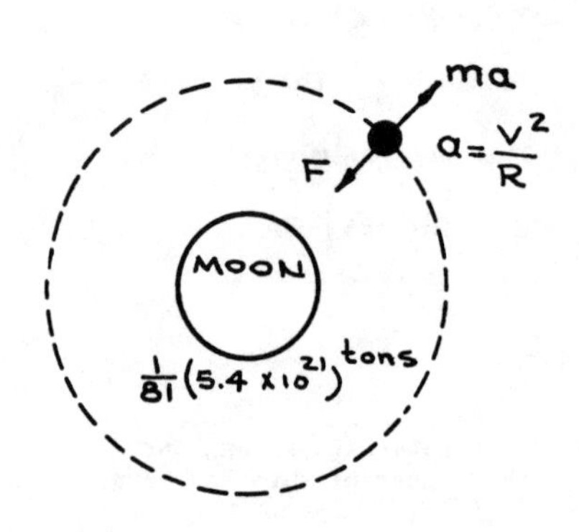

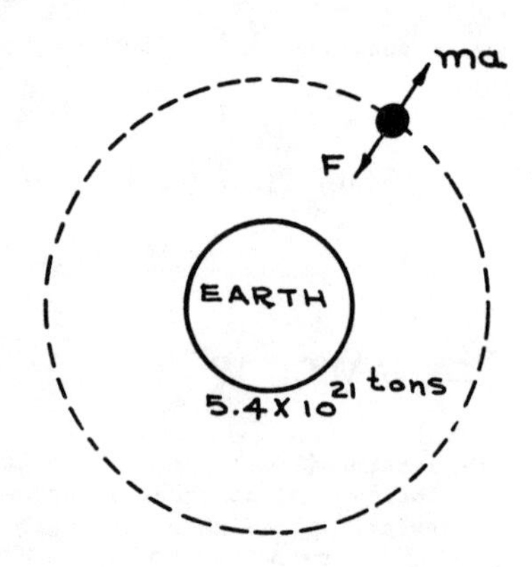

$F = G\frac{M\,m}{R^2}$ where F = gravitational attraction
G = gravitational constant
M and m = masses
R = distance between centers of the two masses

For equilibrium the gravitational attraction must be counterbalanced by the centrifugal force.

$$G\frac{M\,m}{R^2} = \text{mass}(a) = m\frac{V^2}{R}$$

Solving for V^2 $\quad V^2 = \frac{GM}{R}$

Earth Orbit

$$V_e^{\,2} = \frac{GM_e}{R_e} \quad (1)$$

Moon Orbit

$$V_m^{\,2} = \frac{GM_m}{R_m} \quad (2)$$

Dividing $\frac{(2)}{(1)}$

$$\frac{V_m^{\,2}}{V_e^{\,2}} = \frac{\frac{GM_m}{R_m}}{\frac{GM_e}{R_e}} = \frac{R_eM_m}{R_mM_e}$$

$$V_m = \sqrt{\frac{V_e^{\,2}\,R_e\,M_m}{R_m\,M_e}} = \sqrt{\frac{17{,}000^2(4000 + 200)(1)}{(1000 + 200)(81)}}$$

$$= \sqrt{12.5 \times 10^6} = 3{,}530 \text{ mph}$$ ●

DYNAMICS 14

A projectile weighing 100 pounds strikes the concrete wall of a fort with a impact velocity of 1200 feet per second. The projectile comes to rest in 0.01 second, having penetrated the 8-foot thick wall to a distance of 6 feet. What is the average force exerted on the wall by the projectile?

Knowing that the impulse exerted on the wall is equal to the change of momentum of the projectile, one may write

$$\int F\,dt = m(V_1 - V_2)$$

Assuming F is the average force exerted, it can be considered as a constant, thus the equation becomes:

$$Ft = m(V_1 - V_2) \qquad F(0.01) = \frac{100}{32.2}(1200 - 0)$$

$$F = 3.73 \times 10^5 \text{ pounds}$$ ●

DYNAMICS 15

A 10" diameter pulley is belt driven with a net torque of 250 ft-lbs. The ratio of tensions in the tight to slack sides of the belt is 4 to 1. What is the maximum tension in the belt?

(a) 250 lbs
(b) 83 lbs
(c) 800 lbs
(d) 500 lbs
(e) 333 lbs

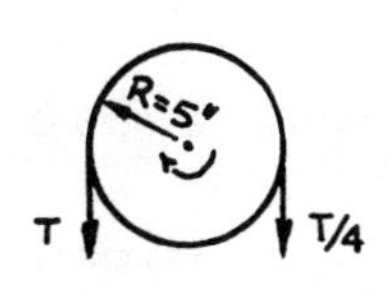

$$\frac{5}{12}T - \frac{5}{12}\cdot\frac{T}{4} = 250 \text{ ft-lbs}$$

$$\frac{3}{4}\cdot\frac{5T}{12} = 250$$

$$T = \frac{250 \times 12 \times 4}{3 \times 5} = 800 \text{ lbs.}$$ ●

Answer is (c)

DYNAMICS 16

Two 3-lb weights are connected by a massless string hanging over a smooth, frictionless peg. If a third weight of 3 lbs is added to one of the weights and the system is released, by how much is the force on the peg increased?

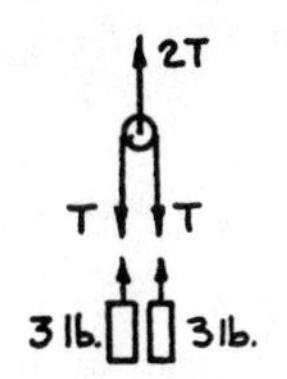

Initial condition:

$T = 3$ lbs.

$2T = 6$ lbs.

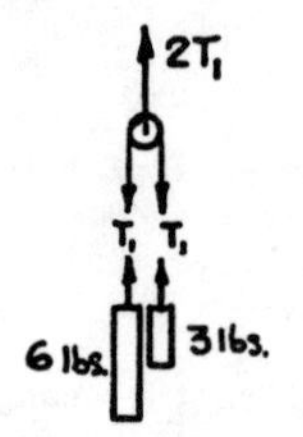

After additional weight added:

$\Sigma F = ma$

$$6 - T_1 = \frac{6}{32.2}a \quad (1)$$

$$T_1 - 3 = \frac{3}{32.2}a \quad (2)$$

Solving these two equations

from (2)
$$a = \frac{32.2}{3}(T_1 - 3)$$

substituting into (1)

$$6 - T_1 = \frac{6}{32.2}\ \frac{32.2}{3}(T_1 - 3)$$

$$6 - T_1 = 2(T_1 - 3) \qquad 3T_1 = 12 \qquad T_1 = 4 \text{ lbs}$$

The force on the peg is increased by the quantity $2T_1 - 2T$

$= 8 - 6 = 2$ lbs. ●

DYNAMICS 17

A body experiences acceleration "a" given by the expression $a = At - Bt^2$, where A and B are constants and t is time. If, at time $t = 0$, the body has zero displacement and velocity, at what next value of time does the body again have zero displacement?

$$a = At - Bt^2 \qquad \frac{dv}{dt} = a \qquad \frac{ds}{dt} = v$$

The differential equation is

$$\frac{dv}{dt} = At - Bt^2 \qquad \text{boundary conditions } t=0,\ v=0.$$

Solving

$$v = A\frac{t^2}{2} - B\frac{t^3}{3} + C$$ where C is a constant of integration

Using the boundary conditions, $v = A(0) - B(0) + C = 0 \qquad \therefore C = 0$

$$v = A\frac{t^2}{2} - B\frac{t^3}{3}$$

The new differential equation is

$$\frac{ds}{dt} = A\frac{t^2}{2} - B\frac{t^3}{3} \qquad \text{boundary conditions } t=0,\ s=0.$$

Solving

$$s = A\frac{t^3}{6} - B\frac{t^4}{12} + D$$ where D is a constant of integration

Using boundary conditions, $s = 0 = A(0) - B(0) + D \qquad \therefore D = 0$

Therefore

$$s = A\frac{t^3}{6} - B\frac{t^4}{12}$$

$$s = \left(\frac{A}{6} - \frac{Bt}{12}\right)t^3$$

Setting $s = 0$

$$s = \left(\frac{A}{6} - \frac{Bt}{12}\right)t^3 = 0$$

$$\frac{A}{6} = \frac{Bt}{12} \qquad t = \frac{2A}{B}$$ ●

DYNAMICS 18

A balloon is rising vertically at the rate of 55 feet per second. When the balloon is 1000 feet above the surface of the earth (flat surface) a bullet is fired from it downward at a 30 degree angle with the vertical. The bullet has a muzzle velocity of 1100 feet per second.

Compute the vertical component of the velocity of the bullet as it strikes the earth.

Note: $g = 32.2$ ft/sec^2. Neglect air friction.

Let the downward direction be positive

$$y = 1000 \text{ feet} \qquad a_y = 32.2 \text{ ft/sec}^2$$

$$\frac{d\,v_y}{dt} = a_y \qquad \therefore v_y = \int a_y dt = a_y t + C$$

imposing the initial conditions

$$v_y(0) = 1100 \cos 30^\circ - 55$$

$$\therefore C = 953 - 55 = 898$$

$$v_y(t) = 898 + a_y t$$

$$y(t) = \int v_y dt = 898t + \frac{a_y t^2}{2} + C_2$$

Measuring from the balloon $y(0) = 0 \quad \therefore C_2 = 0$

At impact with the earth $y = 1000' = 898t + \frac{32.2t^2}{2}$

$$\text{or} \quad 16.1t^2 + 898t - 1000 = 0$$

Solving the quadratic

$$t = \frac{-898 \pm \sqrt{(898)^2 - 4(16.6)(-1000)}}{32.2}$$

$$= \frac{-898 \pm \sqrt{87 \times 10^4}}{32.2} = -27.9 \pm 29.0$$

$t = +1.1$ seconds at impact (the negative root has no physical meaning).

At impact

$$v_y(t) = 898 + 32.2(1.1)$$

$$v_y(1.1) = 933. \text{ feet/second}$$ ●

DYNAMICS 19

If the radius of the earth is approximately 4 times the radius of the moon, and if the mass of the moon is 1/82 of that of the earth, calculate the weight of a pound-mass on the surface of the moon.

The force of attraction between two masses may be computed from the equation:

$$F = G\frac{Mm}{d^2}$$

where G is the gravitational constant
M and m are the masses of the two bodies
d is the distance between the two bodies

1 lbm on the surface of the earth has a weight of 1 lbf, therefore

$$1 = G\frac{M_{earth}(1\ \text{lbm})}{(4R)^2} \qquad (1)$$

Applying this equation now to 1 lbm on the surface of the moon gives:

$$F_{moon} = G\frac{\frac{1}{82}M_{earth}(1\ \text{lbm})}{R^2} \qquad (2)$$

dividing (2) by (1)

$$F_{moon} = \frac{\dfrac{GM_{earth}(1\ \text{lbm})}{82R^2}}{\dfrac{GM_{earth}(1\ \text{lbm})}{16R^2}} = \frac{16}{82} = 0.195\ \text{lbf}$$ ●

DYNAMICS 20

A flywheel weighing 4,480 lbs has a radius of gyration of 3 feet. If there is a driving torque of 600 lb-ft and a resisting torque of 250 lb-ft, find the time required to increase its speed from 20 to 80 revolutions per minute.

Change in angular velocity $= (80 - 20)\frac{2\pi}{60} = 6.28$ radians/second $= \Delta\omega$

Moment of Inertia (I) $= mr^2$ where m = mass
r = radius of gyration (sometimes designated k)

$$I = mr^2 = \frac{W}{g}r^2 = \frac{4480}{32.2}(3)^2 = 1252\ \text{ft-lb-sec}^2$$

Solving by Impulse-Momentum for Rotation:

$$\int_0^t \Sigma M\cdot dt = I\omega - I\omega_o$$

For ΣM = constant, the equation becomes

$$\Sigma Mt = I(\omega - \omega_o)$$

Substituting the values given for this problem

$$(600 - 250)t = 1252(6.28) \qquad t = \frac{1252(6.28)}{350} = 22.5\ \text{seconds}$$ ●

DYNAMICS 21

A pendulum with a concentrated mass (m) is suspended vertically inside a stationary railroad freight car by means of a rigid weightless connecting rod. If the connecting rod is pivoted where it attaches to the box car, compute the angle that the rod makes with the vertical as a result of a constant horizontal acceleration of 2 feet/sec^2 of the box car.

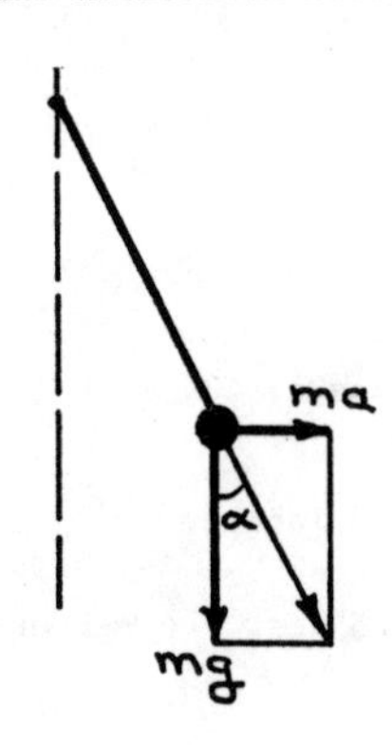

$$\tan \alpha = \frac{ma}{mg} = \frac{a}{g} = \frac{2}{32.2} = 0.0621$$

$$\alpha = 3.55° \; ●$$

DYNAMICS 22

A body moves so that during the first part of its motion its distance S in inches from the starting point is given by the expression

$$S = 6.8t^3 - 10.8t \quad \text{(t in seconds)}$$

The acceleration in 3 seconds would be

(a) 172.8 inches/sec^2
(b) 122.4 " "
(c) 61.2 " "
(d) 212.4 " "
(e) none of the above

$$S = 6.8t^3 - 10.8t$$

$$\frac{ds}{dt} = 3(6.8)t^2 - 10.8 = 20.4t^2 - 10.8$$

$$\frac{d^2s}{dt^2} = 2(20.4)t = 40.8t$$

At t = 3 seconds

$$\frac{d^2s}{dt^2} = 40.8(3) = 122.4 \text{ inches/sec}^2 \; ●$$

Answer is (b)

DYNAMICS 23

A heavy brass plumb-bob suspended from a 38-inch cord was observed to have a natural period of oscillation of about two seconds if pulled 24 inches to one side and then allowed to swing freely. If an astronaut was to repeat this experiment on the moon where the gravitational attraction is approximately 1/6th that of the earth, the observed period would be:

(a) 0.33 seconds
(b) 0.82 "
(c) 4.90 "
(d) 12.00 "
(e) 72.00 "

$T = 2\pi\sqrt{\frac{L}{g}}$ where L = length of pendulum (radius of the arc)
g = acceleration of gravity

The radius of the arc L probably is greater than 38 inches.
Solving for L

$$2 = 2\pi\sqrt{\frac{L}{32.2}} \qquad L = \frac{32.2}{\pi^2} = 3.26 \text{ feet} \quad (39.1 \text{ inches})$$

On the moon

$$g_{moon} = \frac{g}{6}$$

$$T = 2\pi\sqrt{\frac{3.26}{\frac{g}{6}}} = 2\pi\sqrt{\frac{6 \times 3.26}{32.2}} = 4.90 \text{ seconds}$$

Answer is (c)

DYNAMICS 24

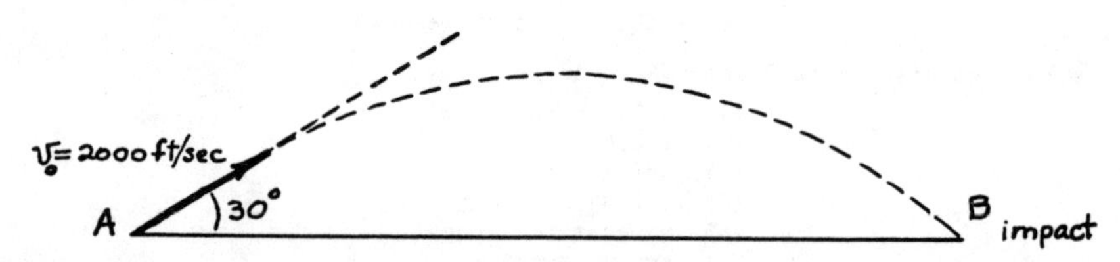

A projectile launched at 2000 ft/sec from A impacts at the same elevation at B. Assume no in-flight propulsion and no aerodynamic drag. Horizontal distance in miles between A and B is nearest to:

(a) 7.9
(b) 10.2
(c) 15.9
(d) 20.3
(e) 24.9

Absence of drag allows horizontal component of velocity to remain at 2000 cos 30° = + 1732 ft/sec throughout flight. Vertical component, initially at 2000 sin 30° = + 1000 ft/sec, is decelerated by gravity to zero vertical velocity at maximum altitude, then accelerated to -1000 ft/sec down at impact. Time of flight is twice the time to reach maximum altitude.

$$v_y = v_{oy} - gt = +1000 - 32.2t = 0 \qquad t = 31 \text{ sec} \qquad 2t = 62 \text{ sec}$$

$$s_x = v_x(2t) = 1732 \times 62.0 = 107{,}000 \text{ ft} = \frac{107{,}000}{5{,}280} = 20.3 \text{ miles}$$

Answer is (d)

4

Mechanics of Materials

MECH OF MTLS 1

Two circular shafts, one hollow and one solid, are made of the same material and have the diameters shown below. If T_h is the twisting moment that the hollow shaft can resist and T_s is the twisting moment that the solid shaft can resist, the ratio of T_h to T_s is:

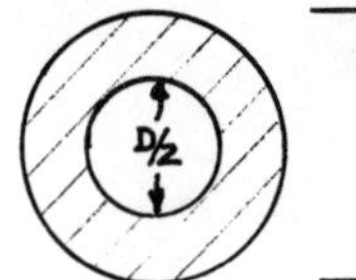

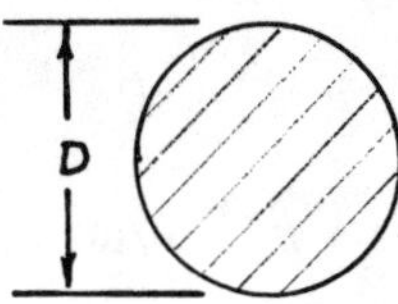

(a) 1/4
(b) 1/2
(c) 9/16
(d) 15/16
(e) 1

Knowing the torsion formula $T = \frac{2S_sJ}{D}$

Compare the polar moments of inertia, J

for circular cross section $J = \frac{\pi D^4}{32}$

for hollow cross section $J = \frac{\pi(D^4 - d^4)}{32} = \frac{\pi}{32}(D^4 - \frac{D^4}{16}) = \frac{\pi D^4}{32}(\frac{15}{16})$

Establish ratio:

$$\frac{T_h}{T_s} = \frac{\frac{2}{D} S_s \frac{\pi D^4}{32} \frac{15}{16}}{\frac{2}{D} S_s \frac{\pi D^4}{32}} = \frac{15}{16}$$ ●

Answer is (d)

MECH OF MTLS 2

A steel cylinder 12 inches outside diameter and a wall thickness of 1 inch is filled with concrete and used as a pier to support an axial load in compression. If the allowable stresses are 40,000 psi for steel and 3,500 psi for concrete and the Youngs modulus for the steel is 30×10^6 psi and for the concrete is 2.8×10^6 psi, what is the allowable load on the pier?

(Assume both materials have linear relationship between stress and strain.)

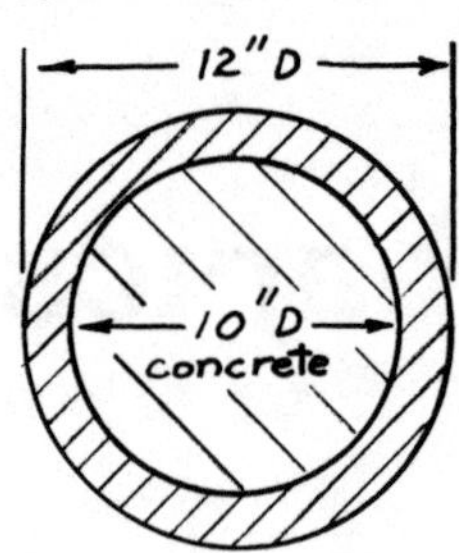

Calculate cross sectional areas

$$\text{concrete:} \quad \frac{\pi}{4} d^2 = \frac{100\pi}{4} = 25\pi \text{ in}^2$$

$$\text{steel:} \quad \frac{\pi}{4}(144) - 25\pi = 11\pi \text{ in}^2$$

In a composite structure maximum allowable strain ϵ of the composite is limited to the lesser of the individual allowable strains.

$$\text{concrete:} \quad \epsilon_{max} = \frac{S_c}{E} = \frac{3.5 \times 10^3}{2.8 \times 10^6} = 1.25 \times 10^{-3} \text{ in/in}$$

$$\text{steel:} \quad \epsilon_{max} = \frac{S_c}{E} = \frac{40 \times 10^3}{30 \times 10^6} = 1.33 \times 10^{-3} \text{ in/in}$$

Both materials are therefore limited to $\epsilon = 1.25 \times 10^{-3}$ in/in and this gives us a reduced allowable steel compressive stress

$$S_c = \frac{1.25 \times 10^{-3}}{1.33 \times 10^{-3}} \times 40{,}000 = 37{,}500 \text{ psi}$$

Total allowable load is the summation of steel and concrete loads (to the allowable ϵ).

Total $P = 3{,}500(25\pi) + 37{,}500(11\pi) = 499{,}500\pi = 1{,}570{,}000$ lb ●

MECH OF MTLS 3

The hollow steel shaft 100 inches long must transmit a torque of 300,000 in-lb. The total angle of twist must not exceed 3 degrees. The maximum shearing stress must not exceed 16,000 psi. Find the Inside Diameter (d) and Outside Diameter (D) of the shaft that meets these conditions.

Shear modulus for steel
$G = 12 \times 10^6$ psi.

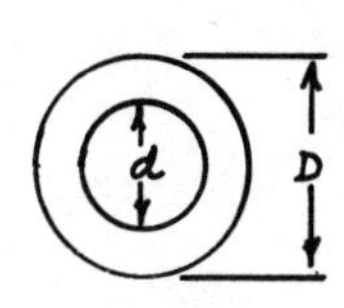

Use torsion formula $T = \frac{2\, S_s\, J}{D}$

Polar Moment of Inertia:

solid shaft: $J = \frac{\pi}{32} D^4$

hollow shaft: $J = \frac{\pi}{32}(D^4 - d^4)$

Shear modulus $G = \frac{S_s}{\phi} = \frac{\text{shear stress, psi}}{\text{shear strain, rad.}}$

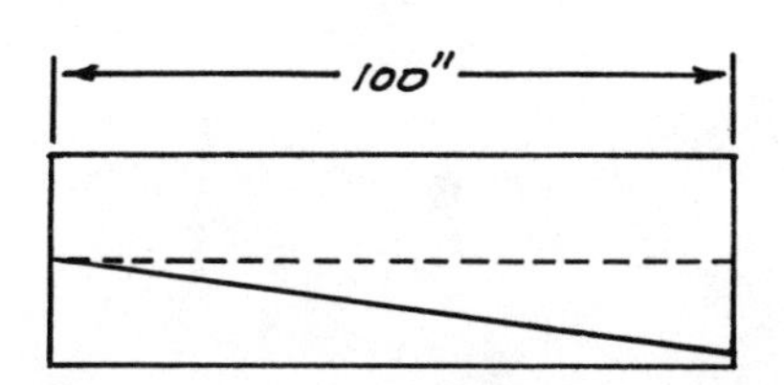

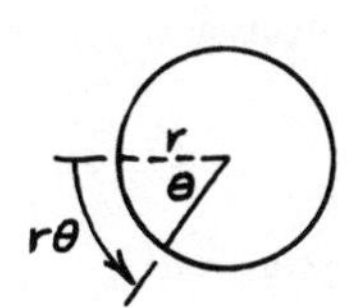

For small angles:

θ in radians = $\sin \theta$

$\theta = 3^o = \frac{3}{57.3}$

$= 0.0524$ rad.

$$\phi = \frac{r\theta}{L} = (\frac{D}{2})\frac{0.0524}{100} = 0.000262\ D$$

Actual shear stress (based on angle of twist):

$$S_s = G\,\phi = 12 \times 10^6 \times 262 \times 10^{-6}\ D = 3{,}144D \text{ psi}$$

Allowable shear stress = 16,000 psi

$$D = \frac{16{,}000}{3{,}144} = 5.10''$$ ● This is twist limited; it may or may not carry the required torque.

Determine that this diameter of <u>solid shaft</u> will transmit the required torque of 300,000 in-lb.

$$T = \frac{2(16{,}000)}{D}\,\frac{\pi}{32}D^4 = 1000\,\pi D^3 = 1000\,\pi(5.10)^3 = 418{,}000 \text{ in-lb.}$$

This is more than adequate, therefore the shaft may be hollowed out for weight reduction.

Use the hollow shaft polar moment of inertia formula in the torsion formula with D = 5.10" and solve for d.

$$300{,}000 = \frac{2(16{,}000)}{D}\,\frac{\pi}{32}(5.10^4 - d^4)$$

$$\frac{300{,}000(5.10)}{1000\,\pi} = 676 - d^4, \qquad d^4 = 188, \qquad d = 3.70''$$ ●

MECH OF MTLS 4

A distance from A to B was measured to be 5,368.25' with a steel tape. The temperature during this time was 22°F. The tape was a standard 100.00' at 68°F. If the coefficient of expansion is 0.0000065, what is the true distance between A and B?

$\Delta L = \alpha L \Delta T$ where α = coefficient of expansion
L = 100 feet
$\Delta T = (22 - 68) = -46°$

$\Delta L = 6.5 \times 10^{-6} \times 100(22 - 68) = -0.0298$ feet

Thus at 22°F the tape was 0.0298 feet shorter than 100 feet.

The tape was used $\frac{5368.25}{100} = 53.6825$ times to measure the distance.

$53.6825 \times 0.0298 = 1.6$ feet

Therefore the distance A to B = 5368.25 - 1.6
= 5366.65 feet ●

MECH OF MTLS 5

The stress in an elastic material is

(a) inversely proportional to the material's yield strength
(b) inversely proportional to the force acting
(c) proportional to the displacement of the material acted upon by the force
(d) inversely proportional to the strain
(e) proportional to the length of the material subject to the force

According to Hooke's law: Stress is directly proportional to strain. Strain is the deformation (or displacement) of the material per unit of length. Thus we can say: Stress is proportional to the displacement of the material acted upon by the force. ●

Answer is (c)

MECH OF MTLS 6

The "SLENDERNESS RATIO" of a column is generally defined as the ratio of its

(a) length to its minimum width
(b) unsupported length to its maximum radius of gyration
(c) length to its moment of inertia
(d) unsupported length to its least radius of gyration
(e) unsupported length to its minimum cross-sectional area

In Euler's equation

$$\sigma_{crit} = \frac{\pi^2 E}{(L/r)^2}$$

where L = unsupported length
r = least radius of gyration

the ratio L/r is called the column slenderness ratio. ●

Answer is (d)

MECH OF MTLS 7

The relationship between the extension of a spring and the force required to produce the extension is

(a) $F = ma$

(b) $F = \mu N$

(c) $F = \frac{mv^2}{R}$

(d) $F = kx$

(e) $F = \mu H$

Hooke's law, that is applicable to springs, is:

$F = kx$ where F = force
k = spring constant
x = change in length

Answer is (d) ●

MECH OF MTLS 8

The linear portion of the stress-strain diagram of steel is known as the

(a) modulus of elongation
(b) plastic range
(c) irreversible range
(d) elastic range
(e) range of yield points

The portion of the diagram where stress is linearly proportional to strain is called the elastic range. ●

Answer is (d)

MECH OF MTLS 9

A spring has a natural length of 12" when at rest. If a force of 10 lbs is required to stretch it 1", find the work required (in ft-lb) to stretch the spring from a total length of 13" to a length of 15".

Using Hooke's law: $F = kx$ where F = force
k = spring constant
x = change of length

If a force of 10 lbs will cause a 1" change of length then

$$k = \frac{F}{x} = \frac{10 \text{ lbs}}{1 \text{ inch}} = 10 \text{ lbs/inch}$$

$$W = \int F\,ds = \int_1^3 10s\,ds = \left.\frac{10s^2}{2}\right|_1^3 = 5(9 - 1) = 40 \text{ inch-lbs}$$

$$= \frac{40 \text{ inch-lbs}}{12 \text{ inchs/ft}} = 3\frac{1}{3} \text{ ft-lbs.}$$ ●

MECH OF MTLS 10

The following compression strengths were recorded for five test cylinders taken from one batch of concrete:

(a) 5,100 psi
(b) 5,620 psi
(c) 5,290 psi
(d) 5,380 psi
(e) 6,110 psi

REQUIRED: Determine the coefficient of variation of these results when the standard deviation, σ, is defined as

$$\sigma = \sqrt{\frac{\sum(x - \bar{x})^2}{n-1}}$$

Standard deviation, σ, is a measure of the dispersion or scatter of a set of values. It is sometimes called the root-mean-square deviation for this is its method of calculation. First, square the deviations of individual values from the arithmetic mean. Then take the mean of these squares and extract the square root.

$$\begin{array}{r} 5{,}100 \\ 5{,}620 \\ 5{,}290 \\ 5{,}380 \\ \underline{6{,}110} \\ \Sigma x = 27{,}500 \end{array} \qquad \text{arithmetic mean} = \frac{27{,}500}{5} = 5{,}500 = \bar{x}$$

	$x - \bar{x}$	$(x - \bar{x})^2$
5,100 - 5,500 =	-400	160,000
5,620 - 5,500 =	120	14,400
5,290 - 5,500 =	-210	44,100
5,380 - 5,500 =	-120	14,400
6,110 - 5,500 =	610	372,100
Σx = 27,500	0	605,000 = $\Sigma(x - \bar{x})^2$

Standard Deviation

$$\sigma = \sqrt{\frac{605{,}000}{4}} = 389$$

$$\text{Coef. of Variation} = \frac{\text{Std Dev}}{\text{mean}} \times 10^2 = \frac{389}{5{,}500} \times 10^2 = 7.07\%$$

MECH OF MTLS 11

Shown on the next page is a frame with the loads and constraints as indicated on the diagram.

REQUIRED: By means of a graphical solution, determine magnitude and direction of the reactions at R_1 and R_2.

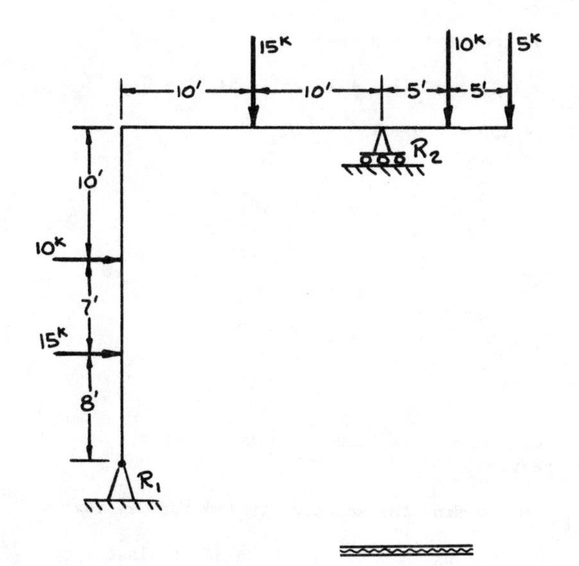
15K
10K
5K
10'
10'
5'
5'
R2
10'
10K
7'
15K
8'
R1

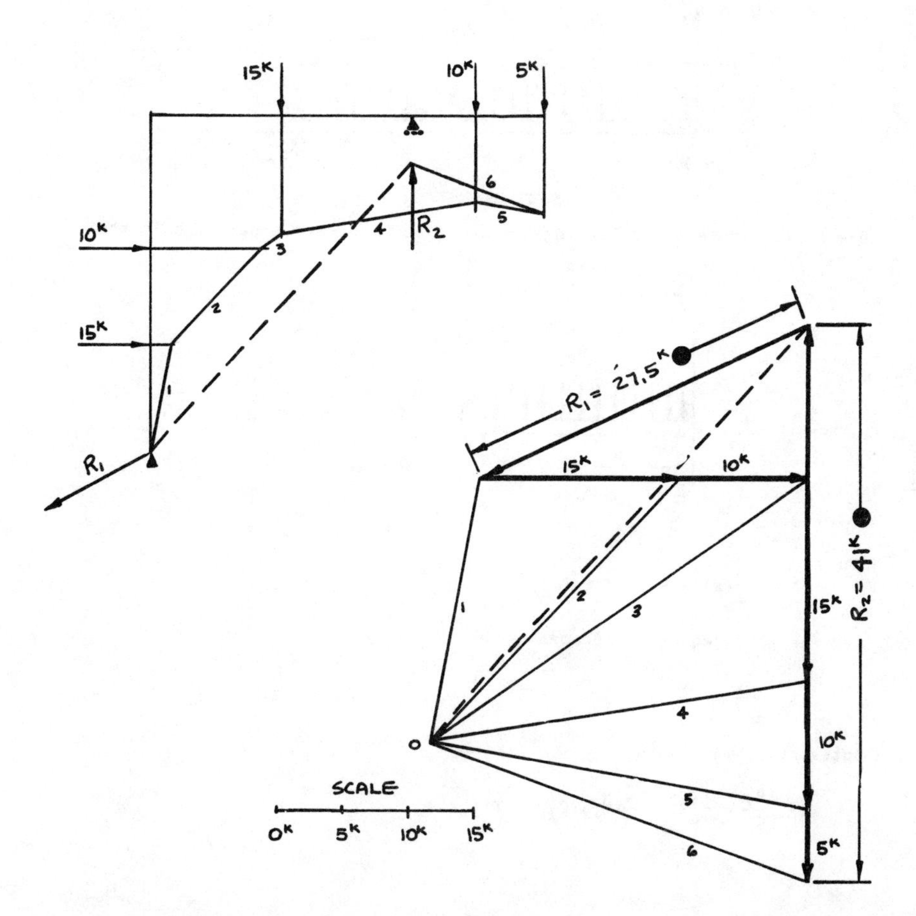
15K
10K
5K
6
5
4
R2
10K
3
2
15K
1
R1
R1 = 27.5K
15K
10K
R2 = 41K
1
2
3
15K
4
O
10K
SCALE
0K
5K
10K
15K
5
6
5K

The analytical solution may be computed as a check.

$$\Sigma M_{R_1} = 0 \quad +15(8) + 10(15) + 15(10) - R_2(20) + 10(25) + 5(30) = 0$$

$$20R_2 = 120 + 150 + 150 + 250 + 150 = 820$$

$$R_2 = \frac{820}{20} = 41^k$$

$$\Sigma F_y = 0 \quad +41 - 15 - 10 - 5 - R_{1_y} = 0 \qquad R_{1_y} = -11^k$$

$$\Sigma F_x = 0 \quad +10 + 15 - R_{1_x} = 0 \qquad R_{1_x} = -25^k$$

$$R_1 = \sqrt{11^2 + 25^2} = 27.3^k$$

MECH OF MTLS 12

Given the beam shown below, with a uniform load per unit length w, a length L, and moment of inertia I.

Solve the equation $EI \frac{d^2y}{dx^2} = M$ to find the equation that describes the deflection of the beam. M = bending moment, E = modulus of elasticity.

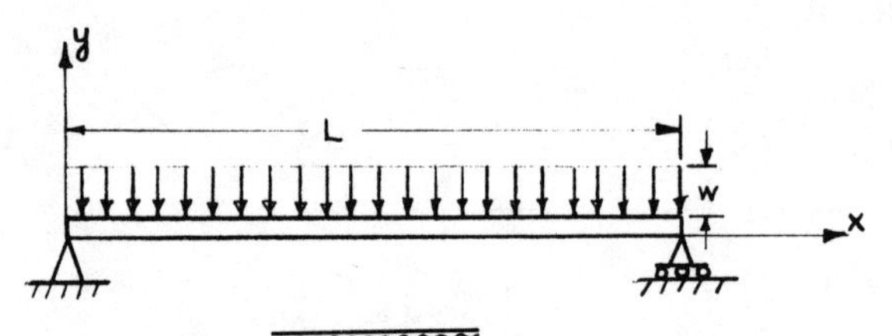

Due to the symmetrical loading the reactions are each equal to half the total load.

$$R_L = R_R = \frac{Lw}{2}$$

Thus

$$M = \frac{Lw}{2}(x) - wx\,\frac{x}{2} = \frac{Lw}{2}x - \frac{w}{2}x^2$$

and the given equation becomes

$$EI \frac{d^2y}{dx^2} = \frac{Lw}{2}x - \frac{w}{2}x^2$$

Integrating we obtain

$$EI \frac{dy}{dx} = \frac{Lw}{4}x^2 - \frac{w}{6}x^3 + C_1 \qquad (1)$$

Because of symmetry the maximum deflection occurs at the midpoint of the span.

Thus $\frac{dy}{dx}$ (slope of the elastic curve) is equal to zero when $x = \frac{L}{2}$

Evaluating C_1

$$0 = \frac{Lw}{4}\frac{L^2}{4} - \frac{w}{6}\frac{L^3}{8} + C_1 \qquad C_1 = -\frac{wL^3}{24}$$

Equation (1) becomes

$$EI\frac{dy}{dx} = \frac{Lw}{4}x^2 - \frac{w}{6}x^3 - \frac{wL^3}{24}$$

Integrating gives us

$$EIy = \frac{wL}{12}x^3 - \frac{w}{24}x^4 - \frac{wL^3}{24}x + C_2 \qquad (2)$$

At $x = 0$ (the support), $y = 0$

Evaluating C_2 all terms are zero and hence $C_2 = 0$

Equation (2) becomes

$$EIy = \frac{wL}{12}x^3 - \frac{w}{24}x^4 - \frac{wL^3}{24}x \qquad (3)$$

Equation (3) is the desired equation. It can be rearranged as follows:

$$y = \frac{wx}{EI}\left(\frac{Lx^2}{12} - \frac{x^3}{24} - \frac{L^3}{24}\right)$$ ●

MECH OF MTLS 13

A steel cylinder surrounded by a copper tube is subject to compression between parallel plates as shown:

$d = 4''$ $\qquad E_s = 30 \times 10^6$ psi

$D = 8''$ $\qquad E_c = 16 \times 10^6$ psi

$P = 100{,}000$ lb

REQUIRED: Determine the stresses in the copper and the steel.

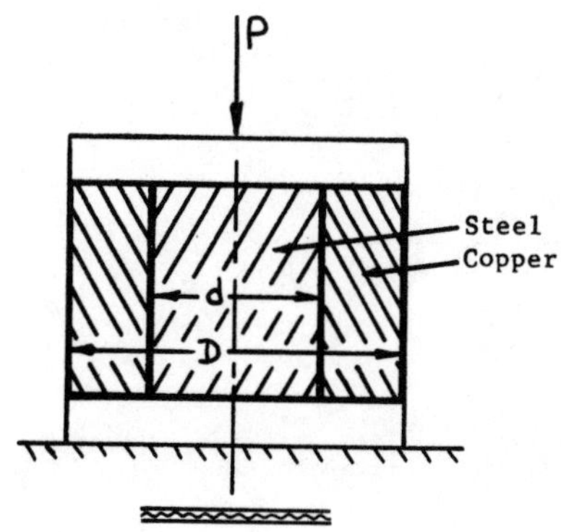

Cross-Sectional Areas

Area Steel

$$A_s = \frac{\pi}{4}d^2 = \frac{\pi}{4}4^2 = 4\pi \text{ in}^2$$

Area Copper

$$A_c = \frac{\pi}{4}D^2 - A_s = \frac{\pi}{4}8^2 - 4\pi = 12\pi \text{ in}^2$$

Since the copper section and the steel section take different loads

$$P = P_c + P_s = 100{,}000 \text{ lbs}$$

From Hooke's law the deformation is

$$\Delta = \frac{PL}{AE}$$ and from the geometry of the problem $\Delta_s = \Delta_c$

Therefore

$$\frac{P_c L}{A_c E_c} = \frac{P_s L}{A_s E_s}$$

Solving for P_c

$$P_c = P_s\left(\frac{A_c E_c}{A_s E_s}\right) = P_s\left(\frac{12\pi}{4\pi}\cdot\frac{16 \times 10^6}{30 \times 10^6}\right) = 1.6\ P_s$$

Thus $100{,}000 = P_c + P_s = 1.6\ P_s + P_s = 2.6\ P_s$

$$P_s = \frac{100{,}000}{2.6} = 38{,}500 \text{ lbs}$$

and $P_c = 100{,}000 - 38{,}500 = 61{,}500$ lbs

$S = \frac{P}{A}$

$$S_s = \frac{P_s}{A_s} = \frac{38{,}500}{4\pi} = 3{,}060 \text{ psi} \bullet$$

$$S_c = \frac{P_c}{A_c} = \frac{61{,}500}{12\pi} = 1{,}630 \text{ psi} \bullet$$

MECH OF MTLS 14

Principal stresses occur on those planes

(a) where the shearing stress is zero
(b) which are 45 degrees apart
(c) where the shearing stress is a maximum
(d) which are subjected only to tension
(e) which are subjected only to compression

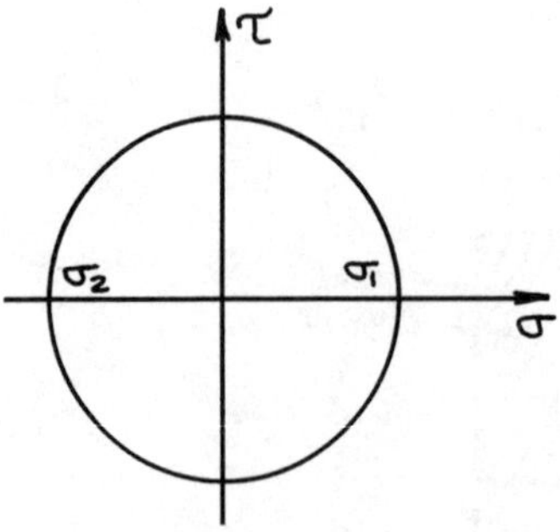

From the Mohr Circle to the left one can see that the shear at the plane of principal stresses is zero. ●

τ

σ_2 σ_1 σ

Answer is (a)

MECH OF MTLS 15

The ratio of the moment of inertia of the cross section of a beam to the section modulus is

(a) equal to the radius of gyration
(b) equal to the area of the cross section
(c) a measure of distance
(d) multiplied by the bending moment to determine the stress
(e) dependent on the modulus of elasticity of the beam material

I = Moment of Inertia
S = Section Modulus = I/c where c is a distance

$$\frac{I}{S} = \frac{I}{I/c} = c$$

Answer is (c) ●

MECH OF MTLS 16

When an air entrainment agent is introduced into a concrete mix

(a) the strength will increase
(b) the strength will decrease
(c) the strength will not be affected
(d) the water/cement ratio must be reduced from 10-15%
(e) calcium chloride should also be added to retard the time of set

Answer is (b) ●

MECH OF MTLS 17

A simple cantilever beam is 10 feet long and has a concentrated load of 100 lbs at its free end. The deflection at the free end is 1".
If the same beam is extended to 20 feet, and the load is reduced to 50 lbs, what will be the deflection at the free end? (Ignore weight of the beam.)

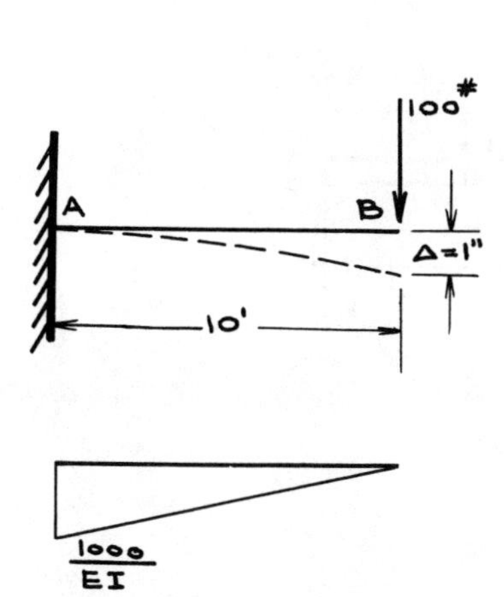

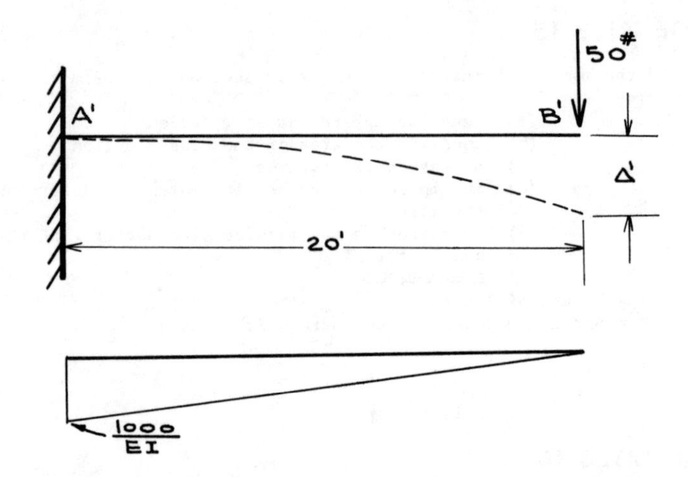

The Second Area-Moment Proposition tells us that the vertical displacement of the free end of a cantilever beam (B) with respect to the fixed end (A) equals the moment (with respect to B) of the area of the bending moment diagram divided by EI.

Or $= \frac{(\text{Area of M diagram})}{EI}\bar{X}$

Thus

$$\Delta = \frac{1000}{EI}(10)(\frac{2}{3} \times 10) \quad \text{and} \quad \Delta' = \frac{1000}{EI}(20)(\frac{2}{3} \times 20)$$

We can see that Δ' is four times that of Δ.

Since Δ = 1", then Δ' = 4" ●

MECH OF MTLS 18

The free body of a certain beam is shown with the values given.
Draw the moment and shear diagrams for this beam, and locate the point of maximum moment.

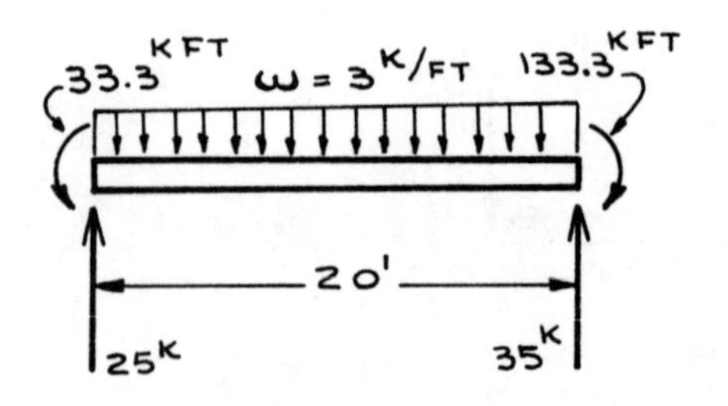

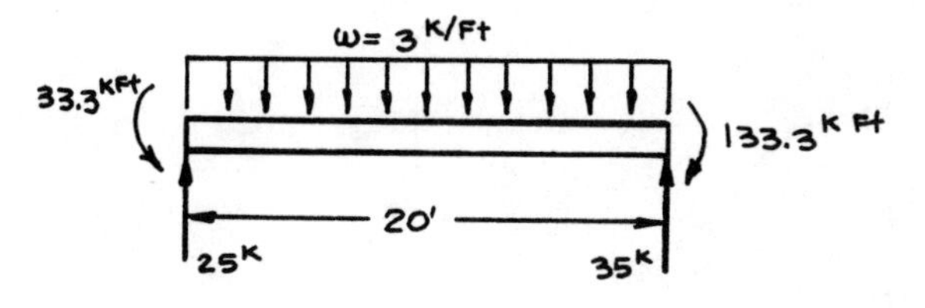

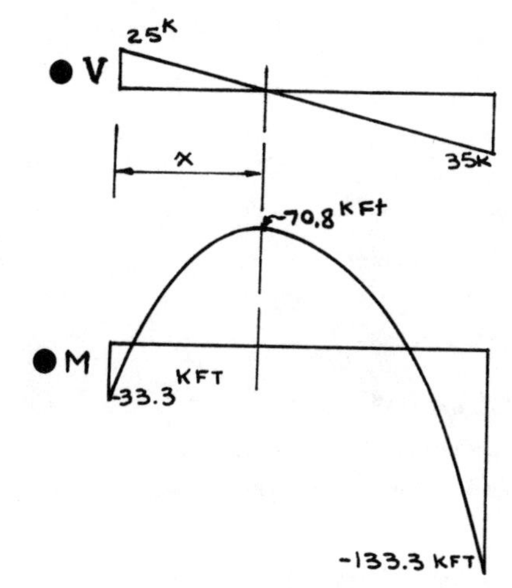

$$X = \frac{25^k}{25^k + 35^k}(20') = 8.33'$$

Maximum Moment
Located at point of zero shear
8.33 feet from left support

$$M_{max} = -33.3 + 25(8.33) - 3^k\left(\frac{8.33^2}{2}\right)$$

$$M_{max} = -33.3 + 208.3 - 104.2$$

$$= 70.8 \text{ k-ft}$$

MECH OF MTLS 19

Four weights of 1, 2, 3, and 4 lbs are attached to a homogeneous wheel of radius 12" as shown. Through what angle θ should the wheel be rotated from the position shown for the weights to

(a) exert a maximum moment about the center of the wheel, and
(b) exert no moment about the center of the wheel?

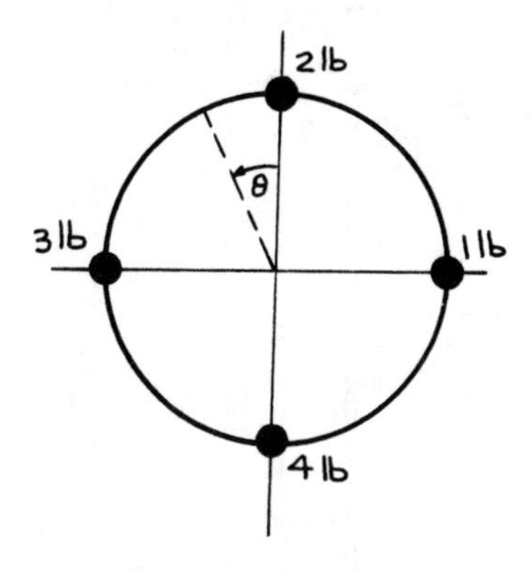

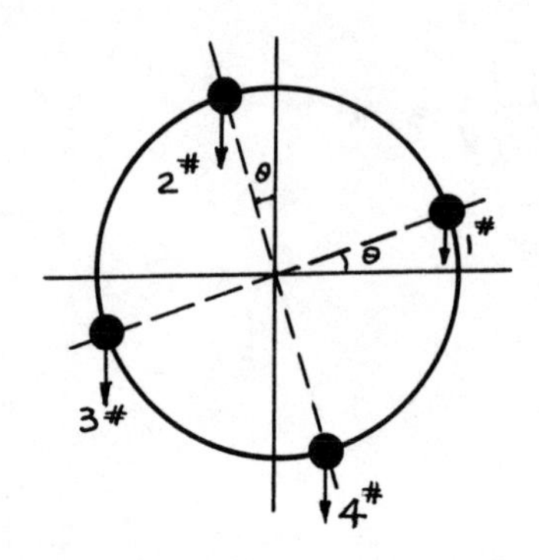

(a) $M = 2(r \sin\theta) + 3(r \cos\theta) - 4(r \sin\theta) - 1(r \cos\theta)$
$\quad = 2(r \cos\theta) - 2(r \sin\theta)$

$$\frac{dM}{d\theta} = 2r(-\sin\theta - \cos\theta) = 0$$

$\sin\theta = -\cos\theta$ $\quad \tan\theta = -1$ $\quad \theta = 135^\circ$ or 315° ●

$\theta = 135^\circ$ gives maximum clockwise moment

$\theta = 315^\circ$ gives max counterclockwise moment

(b)

$M = 2(r \cos\theta) - 2(r \sin\theta) = 0$

$\cos\theta = \sin\theta$ $\quad \tan\theta = 1$ $\quad \theta = 45^\circ$ or 225° ●

MECH OF MTLS 20

A contractor is to construct a water tank, a section of which is shown below, for temporary water supply. The vertical beams are in pairs, as shown, and are spaced 3 feet apart along the entire length of the tank.

Assume that the section shown is near the center of the tank, and neglect end reactions.

Draw the shear and moment diagram, and compute the reactions at the top and bottom of the tank, for one of the vertical beams shown.

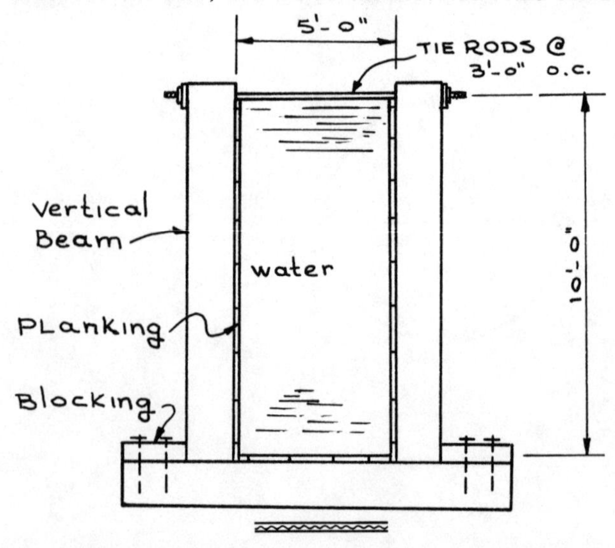

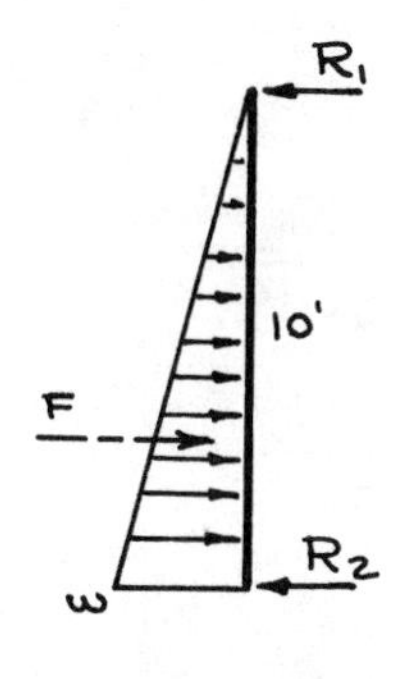

Reactions

$w = 10(62.4)(3') = 1872$ lbs/ft

$F = 1872(\frac{10}{2}) = 9360$ lbs $= 9.36^k$

$\Sigma M_{R1} = 0$

$10R_2 - \frac{2}{3}(10)(9.36) = 0$ $\qquad R_2 = 6.24^k$ ●

$\Sigma F_x = 0$

$6.24 - 9.36 + R_1 = 0$ $\qquad R_1 = 3.12^k$ ●

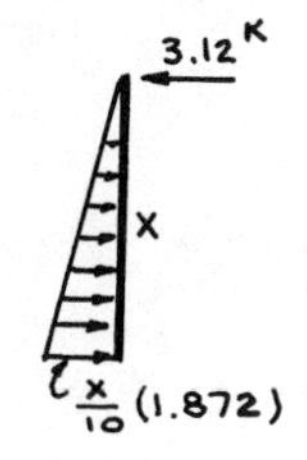

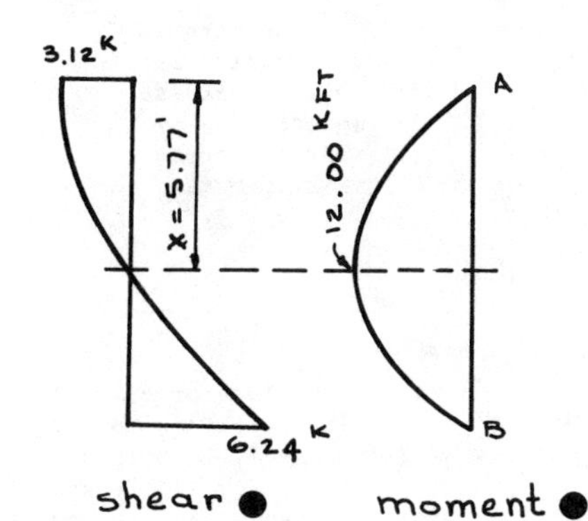

Compute location of zero shear and maximum moment

$\Sigma F_x = 0$ $\qquad 3.12 - \frac{x}{10}(1.872)(\frac{x}{2}) = 0$

$x^2 = \frac{3.12(20)}{1.872} = 33.3$ $\qquad x = 5.77$ feet

Maximum Moment

$M_{max} = 3.12(5.77) - \frac{5.77}{10}(1.872)(\frac{5.77}{2})(\frac{5.77}{3}) = 12.00$ kip-ft

MECH OF MTLS 21

A steam line has the dimensions as shown below at 70°F. What is the total distance the end "P" will move when steam at 450°F is flowing in the line? (Coef. of expansion = 0.0000065 $\frac{ft}{ft^\circ F}$) Assume that all the distances are measured along the principal coordinate (x, y & z) axes.

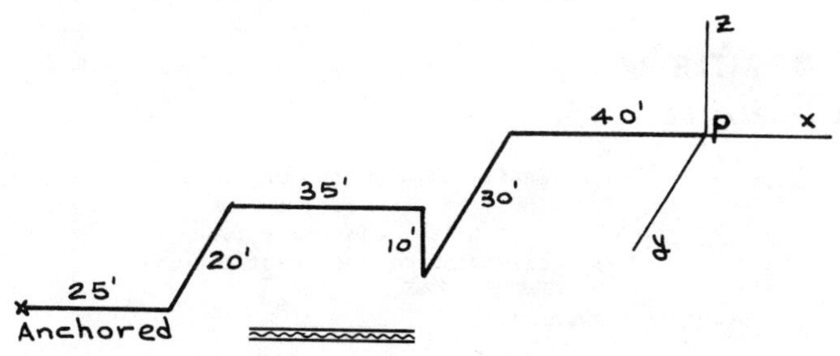

$$x \text{ movement} = (25' + 35' + 40')(450^\circ - 70^\circ)(6.5 \times 10^{-6})$$
$$= (100)(380)(6.5 \times 10^{-6}) = 0.247 \text{ feet}$$

y movement = 50/100 of x movement = 0.123 feet

z movement = 10/100 of x movement = 0.025 feet

$$\text{Resulting movement} = \sqrt{x^2 + y^2 + z^2} = \sqrt{0.247^2 + 0.123^2 + 0.025^2}$$
$$= 0.277 \text{ feet}$$ ●

Note that this solution assumes perfect conditions; frictionless supports.

MECH OF MTLS 22

Structural steel elements subjected to torsion develop

(a) tensile stress
(b) compressive stress
(c) shearing stress
(d) moment
(e) bending stress

Answer is (c) ●

MECH OF MTLS 23

The deflection of a beam is

(a) directly proportional to the modulus of elasticity and moment of inertia
(b) inversely proportional to the modulus of elasticity and length of the beam cubed
(c) inversely proportional to the modulus of elasticity and moment of inertia
(d) directly proportional to the load imposed and inversely to the length squared
(e) inversely proportional to the weight and length

The equation for the second derivative of a deflected beam is

$$\frac{d^2y}{dx^2} = \frac{M}{EI}$$

where y = deflection at any point x from the origin
M = moment at that point
E = modulus of elasticity
I = moment of inertia

So

$$y = \iint \frac{M}{EI}\, dx\,dx$$

Thus the deflection of a beam is inversely proportional to the modulus of elasticity and moment of inertia. ●

Answer is (c)

MECH OF MTLS 24

The differential of the shear equation is which one of the following?

(a) load on the beam
(b) tensile strength of beam
(c) bending moment of the beam
(d) slope of the elastic curve
(e) deflection of the elastic curve

Answer is (a) ●

MECH OF MTLS 25

A homogenous round bar of diameter D, length L, and total weight W is hung vertically from one end. If E is the modulus of elasticity, what is the total elongation of the bar?

(a) $\frac{WL}{\pi D^2 E}$

(b) $\frac{2WL}{\pi D^2 E}$

(c) $\frac{WL}{2\pi D^2 E}$

(d) $\frac{2WLE}{\pi D^2}$

(e) $\frac{WL}{2E}$

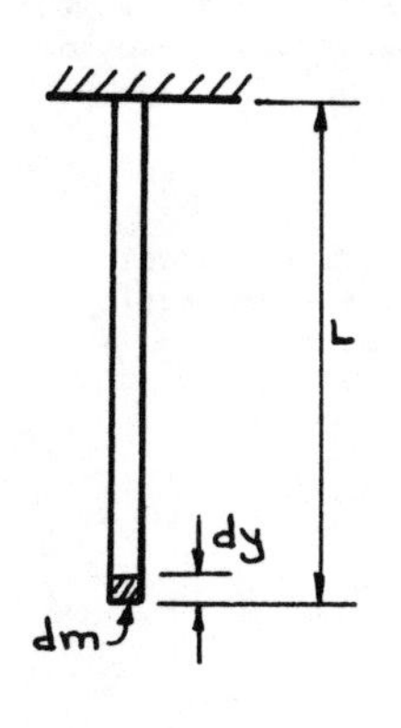

$$dP = dm = dy \cdot \frac{W}{L}$$

$$\Delta = \frac{PL}{AE}$$

$$= \int_0^L \frac{\frac{W}{L}\, dy(L - y)}{AE} = \frac{W}{ELA}\int_0^L (L - y)dy$$

$$= \frac{W}{ELA}\left[-\frac{1}{2}(L - y)^2\right]_0^L = \frac{W}{AEL}\cdot\frac{1}{2}L^2$$

$$= \frac{WL}{2E}\cdot\frac{1}{\frac{\pi}{4}D^2} = \frac{2WL}{\pi D^2 E}$$

Answer is (b)

MECH OF MTLS 26

A vertically loaded beam, fixed at one end and simply supported at the other is indeterminate to what degree?

(a) first
(b) second
(c) third
(d) fourth
(e) statically determinate

There are three unknowns: R_1, R_2, and M and only two equations

$$\Sigma F_v = 0 \quad \text{and} \quad \Sigma M = 0$$

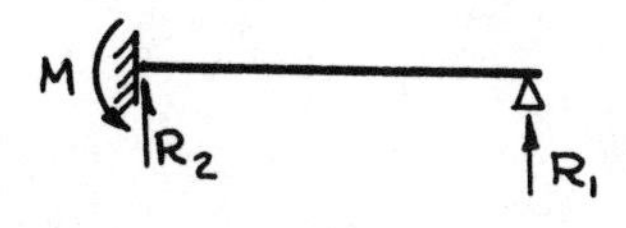

Therefore the beam is indeterminate to the first degree. ●

Answer is (a)

MECH OF MTLS 27

A 60-inch diameter steel pipe, 3/8 inch thick, carries water under a pressure head of 550 feet. Determine the hoop stress in the steel.

Internal Pressure (R) = 550'(62.4 lbs/ft^3) = 34,320 lbs/ft^2 = 238 psi

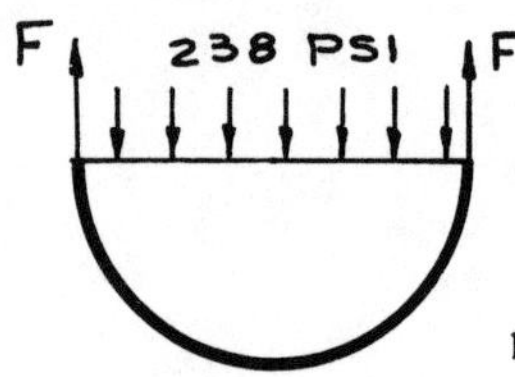

Examine a section of pipe 1" long

2F = RA

= 238 psi (60" x 1")

F = 7,140 lbs

$$\text{hoop stress} = \frac{F}{A} = \frac{7{,}140}{1'' \times \frac{3''}{8}} = 19{,}040 \text{ psi}$$ ●

MECH OF MTLS 28

Given the steel box section shown in the figure.

REQUIRED: (a) Determine the dimensions of a square, solid wood post that has the same column strength. Buckling is not a problem.

Allowable stresses:
wood = 400 psi
steel = 20,000 psi

(b) If wood costs $2.75/cubic foot, installed, and steel $0.30/pound, installed, which is the more economical to use. Steel weighs 490 pounds/cubic foot.

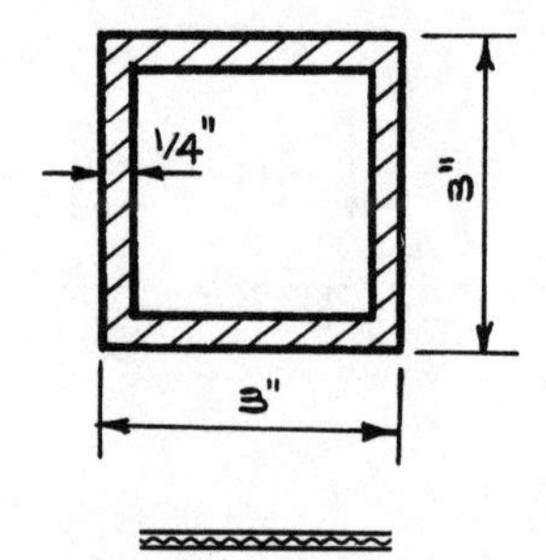

(a) Since buckling is not a problem, the direct stress P/A governs.

$$A_s = 2(3 \times \frac{1}{4}) + 2(2\frac{1}{2} \times \frac{1}{4}) = 2.75 \text{ in}^2$$

$$\text{Required } A_w = \frac{20{,}000}{400} \times 2.75 = 137.5 \text{ in}^2$$

Width of side of square wood post = $(137.5)^{1/2}$ = 11.7 inches ●

(b) Cost/foot of steel column = $\frac{2.75}{144} \times 490 \times 0.30$ = \$2.81

Cost/foot of wood column = $\frac{137.5}{144} \times \$2.75$ = \$2.63

The installed cost of the wood column is less than that of the steel column. ●

MECH OF MTLS 29

A system consisting of two prismatical bars of equal length L and equal cross-sectional area A carries a vertical load P as shown:

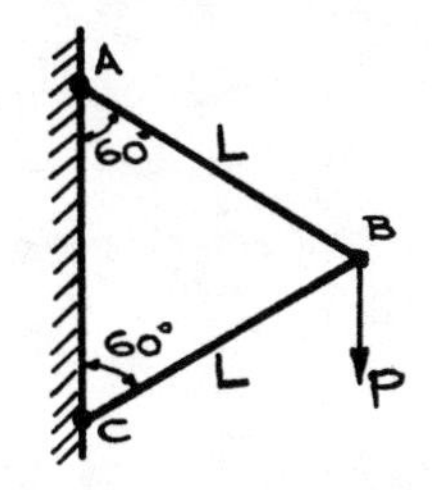

Required: Determine the vertical displacement of hinge B, assuming A and C are hinged points.

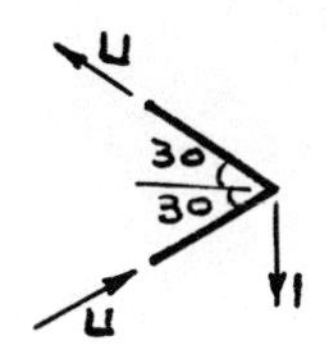

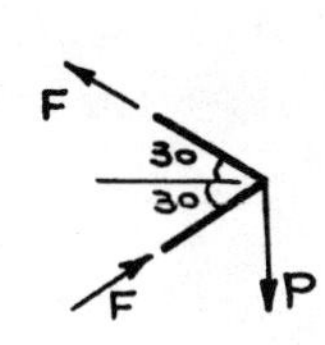

Solve for F

$F \sin 30^\circ + F \sin 30^\circ = P$

$1/2\ F + 1/2\ F = P$

$F = P$

and $u = 1$

Solution by the Method of Virtual Work:

Bar	Length	Area	L/A	Stress due to applied load (S)	Stress due to dummy unit load (u)	$\frac{SLu}{A}$
AB	L	A	L/A	P	1	PL/A
BC	L	A	L/A	P	1	PL/A
						2PL/A

$$\Delta_B = \Sigma\left(\frac{SLu}{AE}\right) = \frac{2PL}{AE}$$

MECH OF MTLS 30

Given an 18 WF 50 beam fixed at one end, pinned at the other end, and externally loaded as shown. Determine the total end moments and reactions (neglect the weight of the beam).

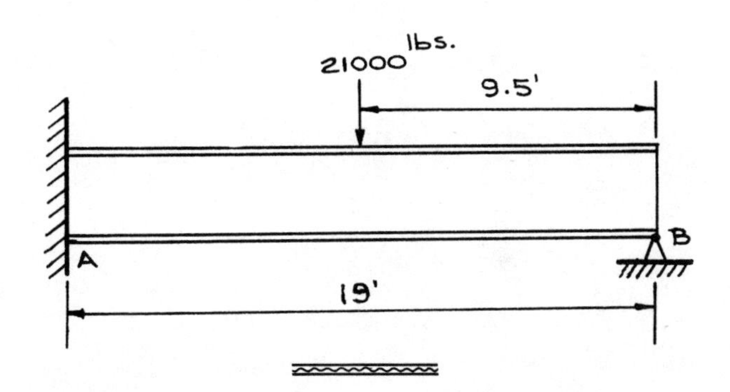

The restrained beam is statically indeterminate.
The solution will be obtained by the Area-Moment Method.

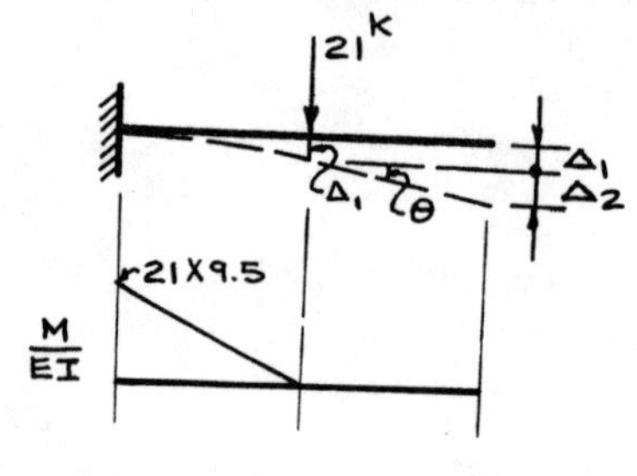

$$\Delta_1 = \frac{1}{EI}(21 \times 9.5)\left(\frac{9.5}{2}\right)\left(\frac{2}{3} \times 9.5\right) = \frac{7 \times 9.5^3}{EI}$$

$$\Delta_2 = 9.5\theta = \frac{9.5}{EI}(21 \times 9.5)\left(\frac{9.5}{2}\right) = \frac{10.5 \times 9.5^3}{EI}$$

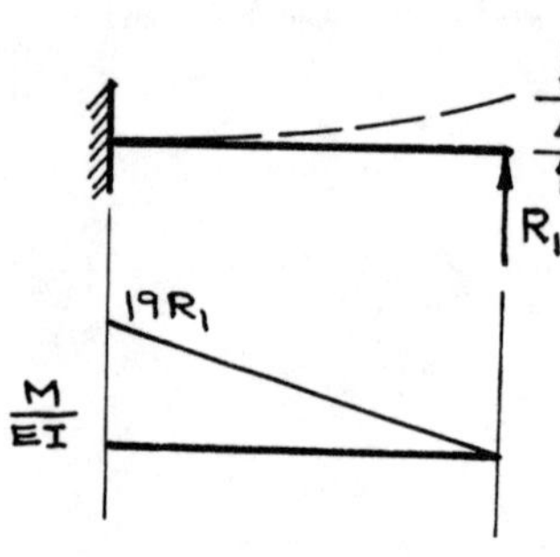

$$\Delta_3 = \frac{1}{EI}(19R_1)\left(\frac{19}{2}\right)\left(\frac{2}{3} \times 19\right) = \frac{R_1}{3EI}(2^3)(9.5^3)$$

$$\Delta_1 + \Delta_2 = \Delta_3$$

$$\frac{7 \times 9.5^3}{EI} + \frac{10.5 \times 9.5^3}{EI} = \frac{8R_1 \times 9.5^3}{3EI}$$

$$7 + 10.5 = \frac{8}{3} R_1$$

$$R_1 = \frac{3 \times 17.5}{8} = 6.56^k$$ ●

M A 9.5' 21K 9.5' B
R2 R1

$$R_2 = 21^k - R_1 = 21^k - 6.56^k = 14.44^k$$ ●

$$M_A = 21^k \times 9.5' - 6.56^k \times 19'$$
$$= 199.5 - 124.6$$
$$= 74.9^{k'}$$ ●

$$M_B = 0$$ ●

MECH OF MTLS 31

Under the loading situation and conditions shown, find the stress in each circular rod due to the 10,000 pound load. The lengths of the 3 rods were the same before the load was applied, and their deformations are equal.

$E_s = 30 \times 10^6$ psi $\qquad E_c = 18 \times 10^6$ psi

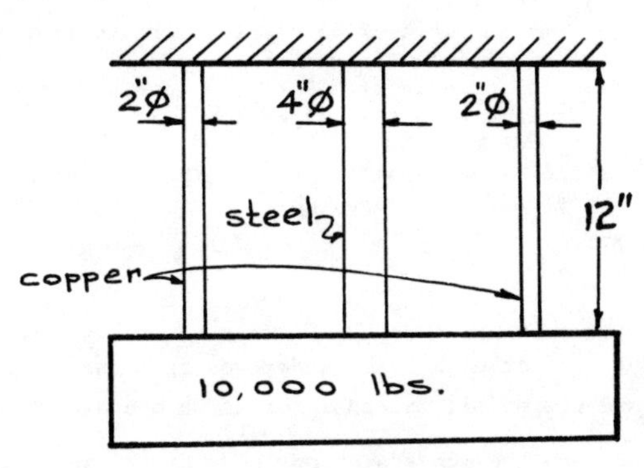

Due to the symmetry the 2 copper rods each carry the same load.

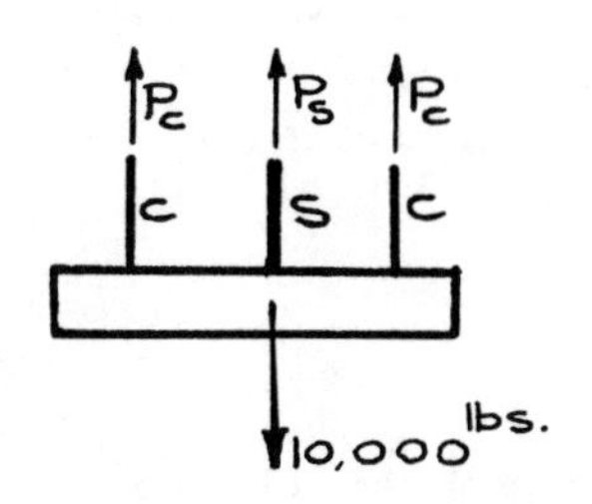

$$2P_c + P_s = 10{,}000 \text{ pounds} \qquad (1)$$

$$\Delta_c = \frac{P_c\ L_c}{A_c\ E_c} = \frac{P_c \times 12}{1^2 \times \pi \times 18 \times 10^6}$$

$$\Delta_s = \frac{P_s\ L_s}{A_s\ E_s} = \frac{P_s \times 12}{2^2 \times \pi \times 30 \times 10^6}$$

$$\Delta_c = \Delta_s$$

$$\frac{12P_c}{18\pi \times 10^6} = \frac{12P_s}{120\pi \times 10^6}$$

$$120P_c = 18P_s \quad \text{or} \quad 20P_c = 3P_s \qquad (2)$$

Solving (1) and (2) gives

$$20P_c = 3(10{,}000 - 2P_c)$$

$$P_c = \frac{30{,}000}{20 + 6} = 1155 \text{ pounds}$$

$$\text{Stress}_c = \frac{1155}{\pi \times 1^2} = 367 \text{ psi} \bullet$$

$$P_s = 10{,}000 - 2(1155) = 7690 \text{ pounds}$$

$$\text{Stress}_s = \frac{7690}{\pi \times 2^2} = 611 \text{ psi} \bullet$$

MECH OF MTLS 32

A thin walled pressurized vessel consists of a right circular cylinder with flat ends. Midway between the ends the stress is greatest in what direction?

(a) Longitudinal
(b) Circumferential
(c) Radial
(d) At an angle of 45° to the longitudinal and circumferential direction.
(e) At an angle of 30° to the longitudinal and circumferential direction.

In thin walled cylinders the circumferential stress ("hoop tension") is twice the value of the longitudinal stress. ●

Answer is (b)

MECH OF MTLS 33

The bending moment at a section of a beam is derived from the

(a) Sum of the moments of all external forces on one side of the section.
(b) Difference between the moments on one side of the section and the opposite side
(c) Sum of the moments of all external forces on both sides of the section
(d) Sum of the moments of all external forces between supports
(e) Difference between the moments on one side and the couple on the other side

The bending moment at a section of a beam is the algebraic sum of the moments of all the external forces on one side of the section. ●

Answer is (a)

MECH OF MTLS 34

The stress concentration factor

(a) Is a ratio of the average stress on a section to the allowable stress
(b) Cannot be evaluated for brittle materials
(c) Is the ratio of areas involved in a sudden change of cross section
(d) Is the ratio of the maximum stress produced in a cross section to the average stress over the section

$$k = \frac{\text{maximum stress}}{\text{average stress}}$$ ●

Answer is (d)

MECH OF MTLS 35

Poisson's Ratio is the ratio of the

(a) Unit lateral deformation to the unit longitudinal deformation
(b) Unit stress to unit strain
(c) Elastic limit to proportional limit
(d) Shear strain to compressive strain
(e) Elastic limit to ultimate strength

$$\text{Poisson's Ratio} = \frac{\text{unit transverse deformation}}{\text{unit axial deformation}}$$ ●

Answer is (a)

MECH OF MTLS 36

Hooke's Law for an isotropic homogeneous medium experiencing one-dimensional stress is

(a) Stress = E(strain)

(b) Strain = E(stress)

(c) $(\text{Force})(\text{area}) = E \frac{(\text{change in length})}{(\text{length})}$

(d) $\frac{\text{Force}}{\text{Area}} = E \frac{(\text{length})}{(\text{change in length})}$

(e) Strain energy = E(internal energy)

Hooke's Law tells us that stress is proportional to strain.

Stress = Constant x Strain

The constant is called the modulus of elasticity (E).
Thus

Stress = E(Strain) ●

Answer is (a)

MECH OF MTLS 37

The modulus of rigidity of a steel member is:

(a) a function of the length and depth.
(b) defined as the unit shear stress divided by the unit shear deformation.
(c) equal to the modulus of elasticity divided by one plus Poisson's ratio.
(d) defined as the length divided by the moment of inertia.
(e) equal to approximately seven-tenths of the modulus of elasticity.

modulus of rigidity = shear modulus, G, psi = $\frac{S_s\text{, psi}}{\phi\text{, radians}}$

$$G = \frac{\text{shear stress}}{\text{shear deformation (strain)}}$$

Answer is (b) ●

MECH OF MTLS 38

A thin homogeneous metallic plate containing a hole is heated sufficiently to cause expansion. If the coefficient of surface expansion is linear, the area of the hole will:

(a) Increase at twice the rate the area of the metal increases
(b) Increase at the same rate as the area of the metal increases
(c) Stay the same
(d) Decrease at the same rate as the area of the metal increases
(e) Decrease at twice the rate the area of the metal increases

Answer is (b) ●

MECH OF MTLS 39

What is the maximum moment of a loaded beam given only the following?

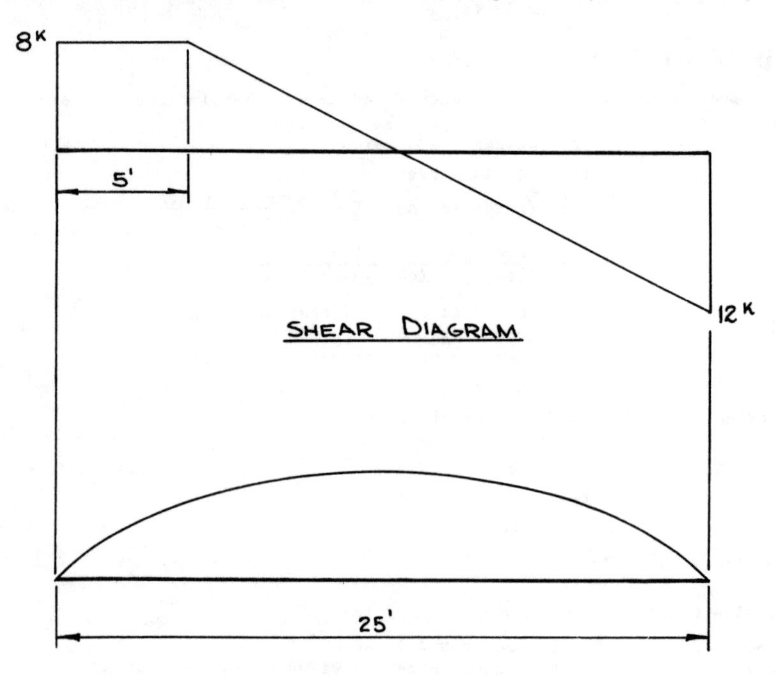

The solution is based on two facts:

1. The maximum value of bending moment occurs where the shear passes through zero.
2. The change in bending moment between two sections of a beam equals the area of the shear diagram between those sections, taking the sign of the shear into account.

The location of zero shear (on the shear diagram) must first be determined.

Its distance from the right end of the beam $= \frac{12}{20}(25'-5') = 12$ feet

The bending moment at this point:

$$M_{max} = 12^k \times \frac{12'}{2} = +72 \text{ kip-feet}$$

MECH OF MTLS 40

A steel bar is mounted as shown, between unyielding supports. If the temperature of the bar increases by ΔT, find the unit stress in the bar in terms of the coefficient of thermal expansion α and the modulus of elasticity E.

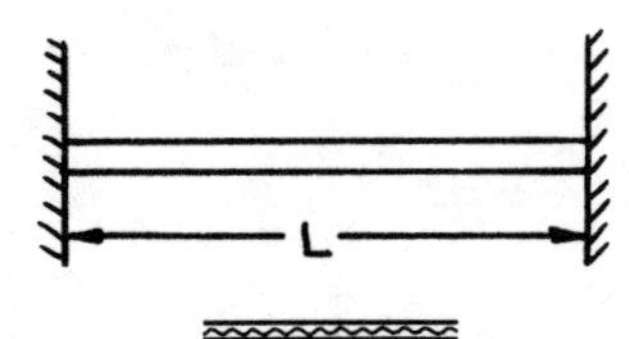

α = coefficient of thermal expansion

E = modulus of elasticity

ΔT = increase in termperature

Strain $\delta = \alpha(\Delta T)$

Since Stress $S = E\delta$ Then Stress $S = E\alpha(\Delta T)$ ●

Note that in a body in which all temperature deformation is prevented, the unit stress caused by thermal expansion is independent of the length L of the body.

MECH OF MTLS 41

A beam of uniform cross-section, symmetrical about its center, is subjected to pure bending. The fiber stresses in the beam vary from zero in the center to a maximum in the extreme fibers, and the resisting moment varies in a similar manner. I-beams and wide flange beams are two types of beams designed to take advantage of this fact in order to save weight.

If the rectangular beam shown is used, how many times the resisting moment of area A (the middle two quarters) would areas B (outer two quarters) have?

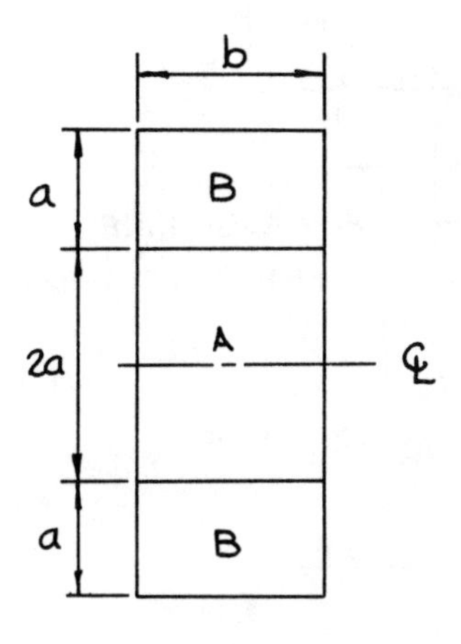

The flecture formula

$$f_s = \frac{Mc}{I}$$

$M_{resisting} = \frac{f_s I}{c}$ For a given section $M_{resisting} \propto I$

$I_A = \frac{bd^3}{12} = \frac{b(2a)^3}{12} = \frac{8ba^3}{12}$ $I_{A+B} = \frac{bd^3}{12} = \frac{b(4a)^3}{12} = \frac{64ba^3}{12}$

$I_B = I_{A+B} - I_A = \frac{64ba^3}{12} - \frac{8ba^3}{12} = \frac{56ba^3}{12}$

$$\frac{M_{resisting(B)}}{M_{resisting(A)}} \propto \frac{I_B}{I_A} = \frac{\frac{56ba^3}{12}}{\frac{8ba^3}{12}} = 7$$ ●

MECH OF MTLS 42

A steel structural ∠6"x6"x1/2" must be welded to the top of a steel WF beam. The design weld value for the weld shown is 3,600 lb/lin. inch.

If the tension load on the angle is 20,000 pounds, show the length and placing of welds so that no eccentricity will exist in the connection.

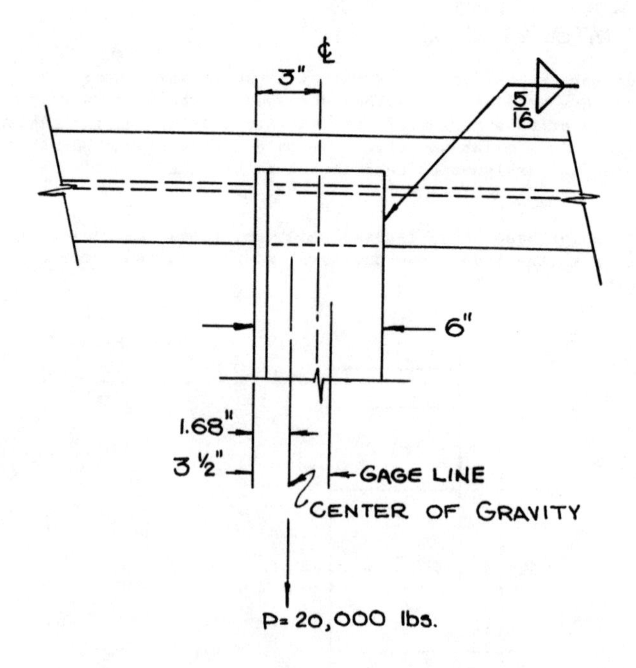

PLAN VIEW

Required length of weld = $\frac{20,000}{3,600}$ = 5.57 inches

Let X = Length of weld on the side with the outstanding leg
$5.57 - X$ = Length of weld on the right side of the angle

For no eccentricity the centroid of the welds must be at the center of gravity of the angle, or

$$1.68X = (6 - 1.68)(5.57 - X)$$
$$= 24 - 4.32X$$
$$6X = 24$$

$X = 4.00$ inches ●
$5.57 - X = 1.57$ inches ●

MECH OF MTLS 43

Determine the deflection of the uniformly loaded beam with simply supported ends, as a function of X. Let W be the load per unit length, L the length, E the modulus of elasticity, and I the moment of inertia.

Given: $M = EI\frac{d^2y}{dx^2}$

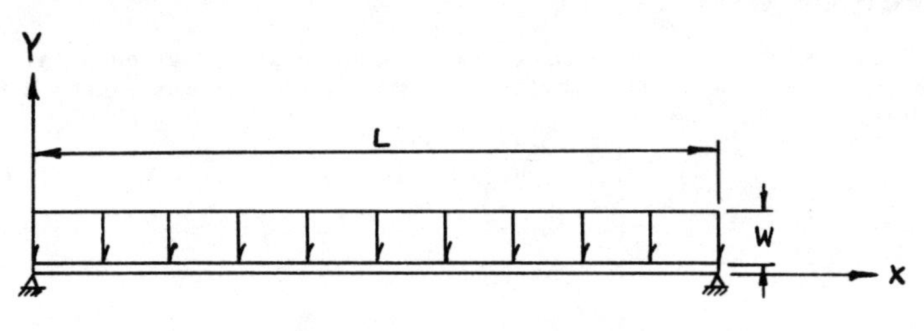

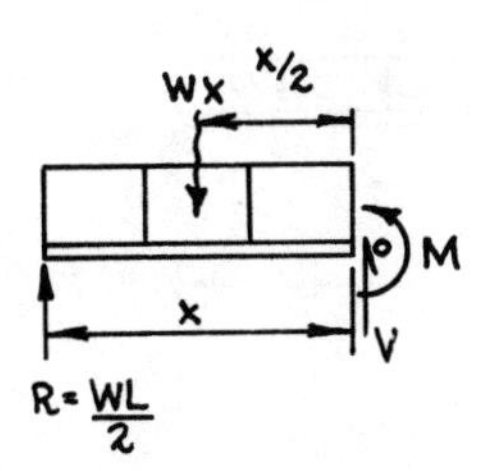

$M = EI\frac{d^2y}{dx^2}$

$\Sigma M_o = 0$ (+)

$$\frac{WL}{2}(X) - WX(\frac{X}{2}) - M = 0$$

$$M = \frac{WLX}{2} - \frac{WX^2}{2}$$

Successive integrations give:

$$EI\frac{dy}{dx} = \frac{WLX^2}{4} - \frac{WX^3}{6} + C_1 \quad (1)$$

$$EIy = \frac{WLX^3}{12} - \frac{WX^4}{24} + C_1X + C_2 \quad (2)$$

Boundary conditions:

$$\frac{dy}{dx} = 0 \qquad \text{when} \quad X = \frac{L}{2}$$

$$y = 0 \qquad \text{when} \quad X = 0$$

Substituting the boundary conditions into (1) and (2)

$$EI(0) = \frac{WL(L/2)^2}{4} - \frac{W(L/2)^3}{6} + C_1 \qquad C_1 = -\frac{WL^3}{24}$$

$$EI(0) = \frac{WL(0)^3}{12} - \frac{W(0)^4}{24} - \frac{WL^3}{24}(0) + C_2 \qquad C_2 = 0$$

Thus equation (2) becomes

$$EIy = \frac{WLX^3}{12} - \frac{WX^4}{24} - \frac{WL^3X}{24}$$

$$\text{Deflection } y = \frac{2WLX^3 - WX^4 - WL^3X}{24EI}$$ ●

MECH OF MTLS 44

The maximum bending moment of a beam simply supported at both ends and subject to a total load W uniformly distributed over its length L is expressed by the formula

(a) $WL/8$
(b) $WL^2/8$
(c) $WL/2$
(d) $WL^2/2$
(e) $WL/4$

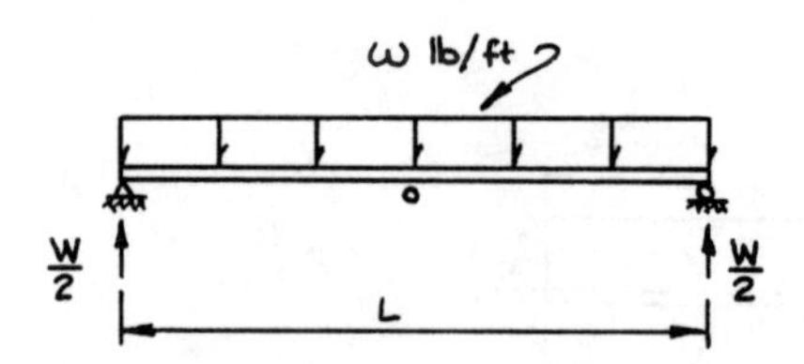

W = ωL

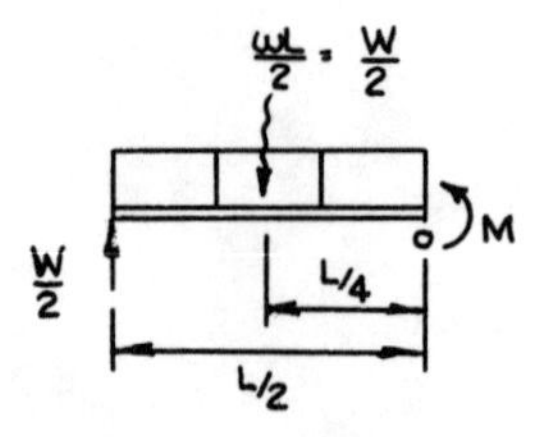

$\Sigma M_o = 0$ (+↺)

$$\left(\frac{W}{2} \times \frac{L}{2}\right) - \left(\frac{W}{2} \times \frac{L}{4}\right) - M = 0$$

$$M = \frac{WL}{4} - \frac{WL}{8} = \frac{WL}{8}$$ ●

Answer is (a)

MECH OF MTLS 45

The maximum moment induced in a simply supported beam of 20 foot span, by a 2,000 pound load at midspan is

(a) 125,000 ft-lbs
(b) 15,000 ft-lbs
(c) 30,000 ft-lbs
(d) 10,000 ft-lbs
(e) 7,500 ft-lbs

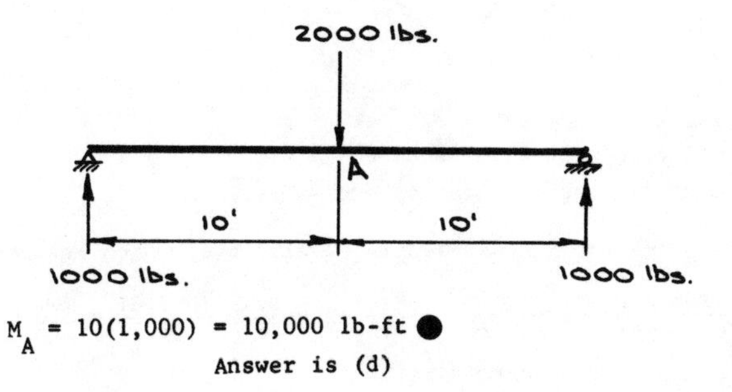

$M_A = 10(1,000) = 10,000$ lb-ft ●

Answer is (d)

MECH OF MTLS 46

The three moment equation may be used to analyze

(a) a continuous beam.
(b) a beam loaded at the third points.
(c) a step tapered column.
(d) a three element composite beam.
(e) an axially end-loaded beam.

Answer is (a) ●

MECH OF MTLS 47

A thin hollow sphere of radius 10" and thickness 0.1" is subjected to an internal pressure of 100 psig. The maximum normal stress on an element of the sphere is

(a) 5,000 psi
(b) 7,070 psi
(c) 10,000 psi
(d) 14,140 psi
(e) 20,000 psi

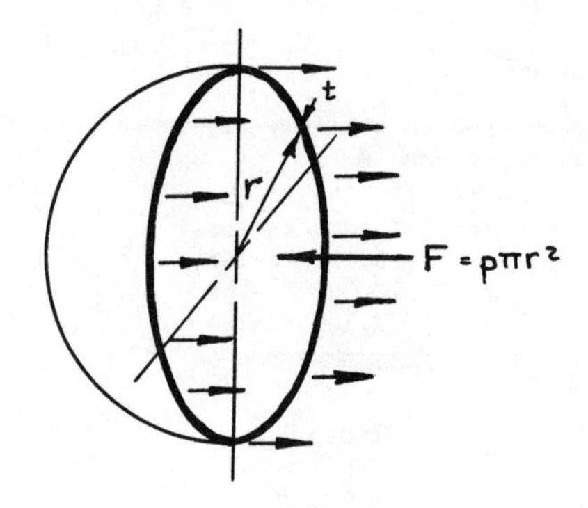

Unit pressure = p
Cross-sectional area of wall of sphere = $2\pi rt$
Area of cross-section through the sphere = πr^2

Force F = unit pressure x cross-sectional area over which it acts = $p\pi r^2$

$$\text{Stress} = \frac{\text{Force F}}{\text{Area of wall of sphere}} = \frac{p\pi r^2}{2\pi rt} = \frac{pr}{2t} = \frac{100(10)}{2(0.1)} = 5{,}000 \text{ psi}$$ ●

Answer is (a)

MECH OF MTLS 48

A tension member made up of two angles carries a load of 160 kips as shown. The angles must be welded to a 7/8" gusset plate with the 4" leg of the angles outstanding. If the value of the weld is 3000 lbs/inch, design the lengths and placement of the welds to suit the given condition.

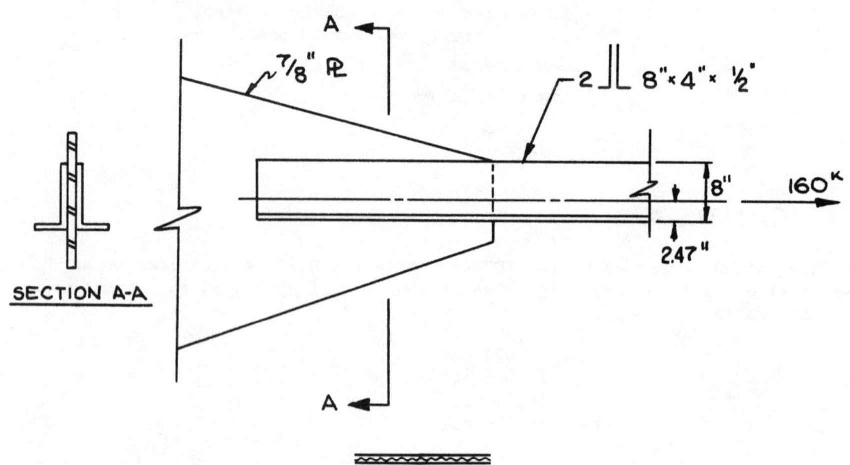

Load per angle = 80^k

$$\text{Inches of weld per angle} = \frac{80^k}{3^k/\text{inch}} = 26.7 \text{ inches}$$

We want the resultant force exerted by the welds to be collinear with the 80 kip force at the centroid of the angle.

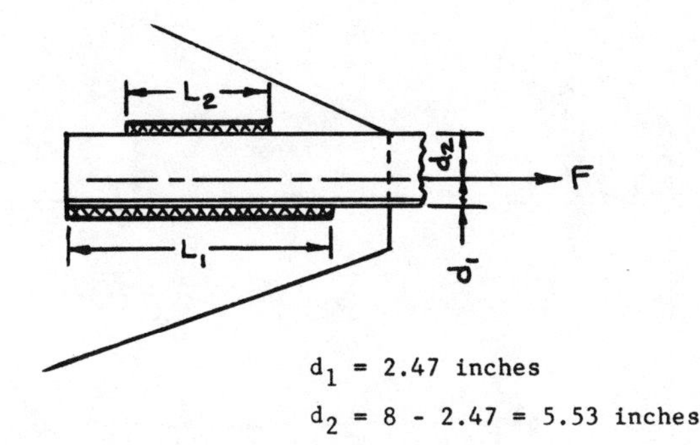

$d_1 = 2.47$ inches

$d_2 = 8 - 2.47 = 5.53$ inches

$$2.47L_1 - 5.53L_2 = 0$$

$$L_1 + L_2 = 26.7$$

Solving the two equations:

$$L_1 = 26.7 - L_2$$

$$2.47(26.7 - L_2) - 5.53L_2 = 0$$

$$66 - 2.47L_2 - 5.53L_2 = 0$$

$$-8L_2 = -66 \qquad L_2 = 8.25 \text{ inches} ●$$

$$L_1 = 26.7 - 8.25 \qquad L_1 = 18.45 \text{ inches} ●$$

MECH OF MTLS 49

Calculate the reactions at A and B and draw the shear and moment diagrams. Neglect the weight of the beam.

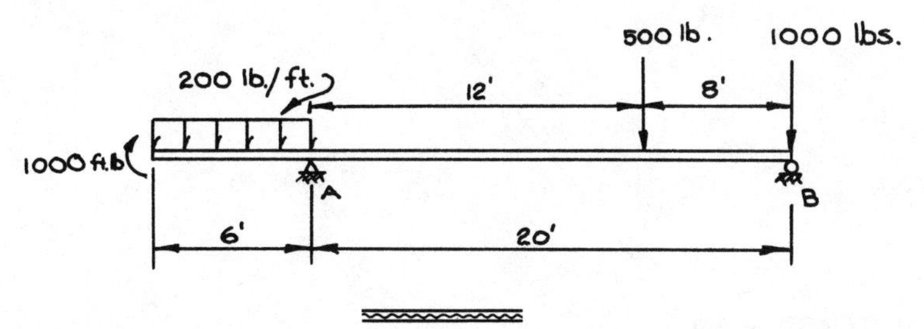

$$\Sigma M_B = 0 \qquad -8(500) + 20R_A - 23(200 \times 6) + 1{,}000 = 0$$

$$20R_A = 4000 + 23(1{,}200) - 1{,}000$$

$$R_A = \frac{3{,}000 + 27{,}600}{20} = 1{,}530 \text{ lbs} ●$$

$$\Sigma M_A = 0 \qquad 1{,}000 - 3(200 \times 6) + 12(500) + 20(1{,}000) - 20R_B = 0$$

$$20R_B = 1{,}000 - 3{,}600 + 6{,}000 + 20{,}000$$

$$R_B = \frac{23{,}400}{20} = 1{,}170 \text{ lbs} ●$$

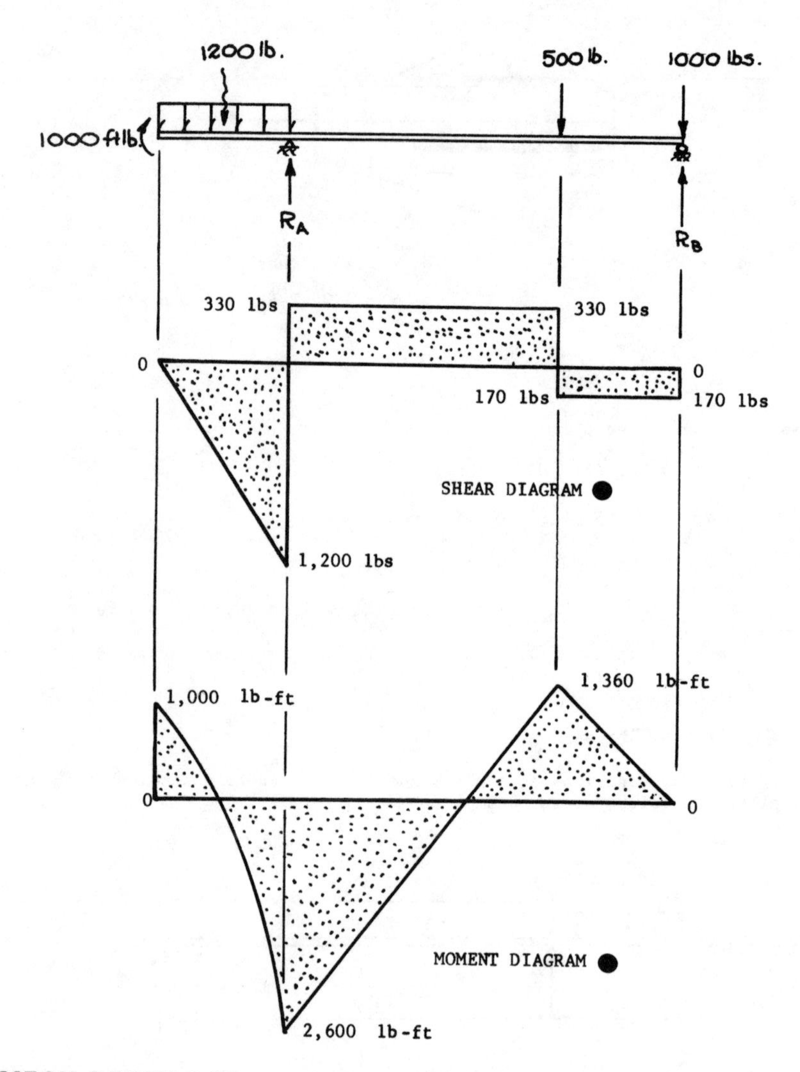

MECH OF MTLS 50

Force F, and moment M, act at point P. Both bars have length L, Area A, modulus of elasticity E, and moment of inertia I.

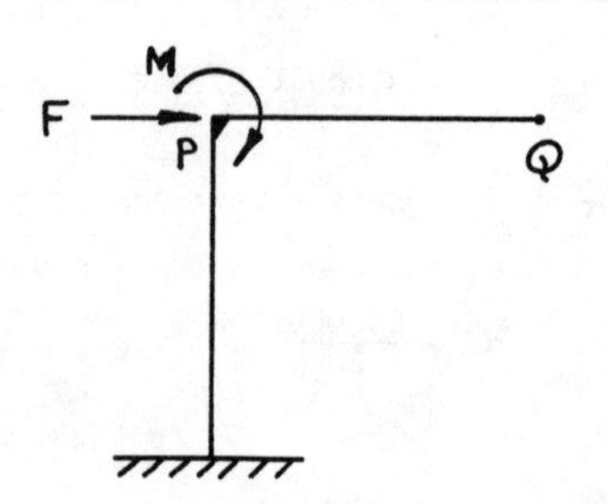

(a) What is the vertical deflection of Q in terms of the parameters given above?
(b) What is the horizontal deflection of Q in terms of the parameters given above?

Solution by the conjugate beam method:

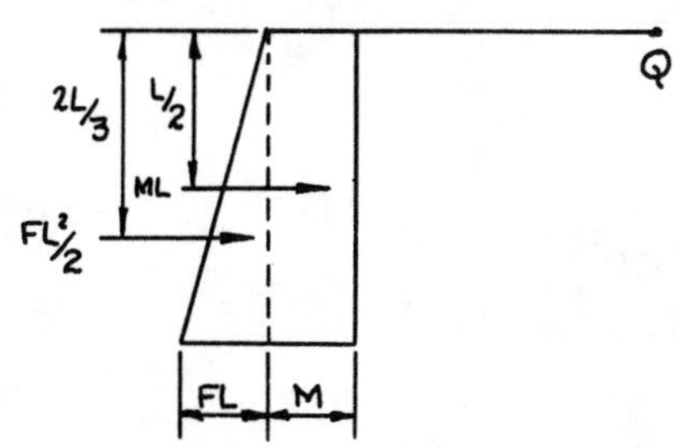

MOMENT DIAGRAM

$$Q_H = \left[\left(\frac{FL^2}{2}\right)\left(\frac{2L}{3}\right) + (ML)\left(\frac{L}{2}\right)\right]\frac{1}{EI} = \frac{FL^3}{3EI} + \frac{ML^2}{2EI} \rightarrow$$

$$Q_V = \left[\left(\frac{FL^2}{2}\right)(L) + (ML)(L)\right]\frac{1}{EI} = \frac{FL^3}{2EI} + \frac{ML^2}{EI} \downarrow$$

MECH OF MTLS 51

Determine the maximum tensile bending stresses and compressive bending stresses in the cantilever T-beam shown below.

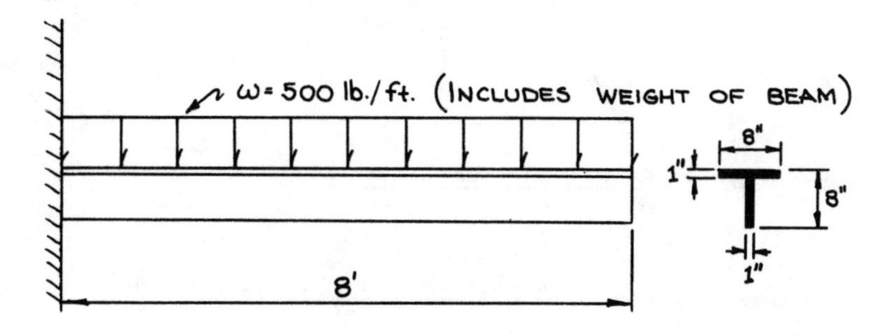

$\text{Stress} = \frac{Mc}{I}$

The maximum stresses will occur at the cross-section where the moment M is a maximum. For this cantilever the maximum moment is at the face of the wall

M_{max} = (500 lb/ft)(8 ft)(4 ft) = 16,000 lb-ft

$$\bar{X} = \frac{\Sigma A\bar{X}}{\Sigma A} = \frac{(8)(1)(7.5) + (7)(1)(3.5)}{15} = \frac{84.5}{15} = 5.63 \text{ inches}$$

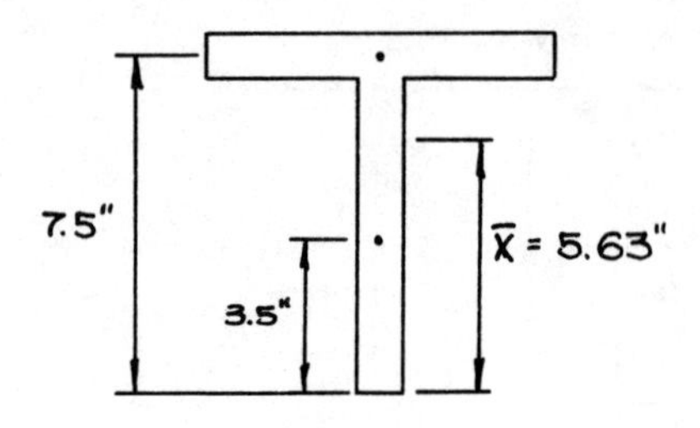

Moment of inertia for composite area:

For rectangular section: $I_o = \frac{bd^3}{12}$

$$I_x = I_o + Ad^2$$

$$I_x = \frac{8 \times 1^3}{12} + 8(7.5 - 5.63)^2 + \frac{1 \times 7^3}{12} + 7(5.63 - 3.5)^2$$

$$= 0.67 + 28.0 + 28.58 + 31.75 = 89.0 \text{ in}^4$$

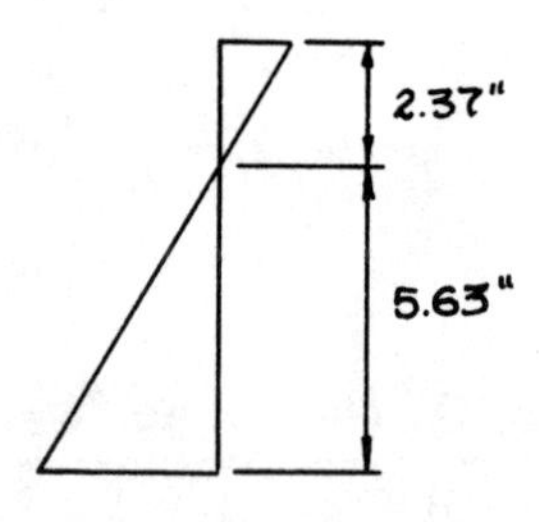

BENDING STRESS DIAGRAM

$$\text{Maximum compressive stress} = \frac{Mc}{I} = \frac{(16{,}000 \text{ ft-lb})(12 \text{ in/ft})(5.63 \text{ in})}{89.0}$$

$$= 12{,}140 \text{ psi}$$ ●

$$\text{Maximum tensile stress} = \frac{Mc}{I} = \frac{(16{,}000 \text{ ft-lb})(12 \text{ in/ft})(2.37 \text{ in})}{89.0}$$

$$= 5{,}120 \text{ psi}$$ ●

MECH OF MTLS 52

A 20" long bimetallic strip is made of metal A and metal B, each 0.020 inches thick. Assuming that the strip is straight at a temperature of 68°F, and is fixed at one end and free to move at the other, compute the deflection of the free end at a temperature of 86°F, if the curvature of the strip is in the form of a circle.

The linear coefficient of expansion for metal A is 24×10^{-6} inches/inch/°C

The linear coefficient of expansion for metal B is 14×10^{-6} inches/inch/°C

Assume that both metals have the same modulus of elasticity.

Determine the temperature change in degrees Centigrade:

$$°C = \frac{5}{9}(F - 32) \qquad °C = \frac{5}{9}(68 - 32) = 20°C$$

$$°C = \frac{5}{9}(86 - 32) = 30°C$$

Difference in expansion between A and B

$$= (24 \times 10^{-6} - 14 \times 10^{-6})(30°C - 20°C)(20 \text{ inches}) = 0.002$$

The shape assumed by the bimetallic strip will be a circular segment.

The circumference of a circle is $2\pi R$

The circumferential length of the circular segment (S) = $R\Theta$ for any angle Θ (in radians).

Let L = heated length of metal B

R = radius to centerline of the bimetallic strip

Substituting into $S = R\Theta$

$$\text{Metal B} \qquad L = (R - 0.01)\Theta \qquad (1)$$

$$\text{Metal A} \qquad L + 0.002 = (R + 0.01)\Theta \qquad (2)$$

Substituting the value of L from (1) into (2)

$$(R - 0.01)\Theta + 0.002 = (R + 0.01)\Theta$$

$$R\Theta - 0.01\Theta + 0.002 = R\Theta + 0.01\Theta$$

$$\Theta = \frac{0.002}{0.02} = 0.1 \text{ radian} = 5.73 \text{ degrees}$$

$$R = \frac{S}{\Theta} = \frac{20}{0.1} = 200 \text{ inches}$$

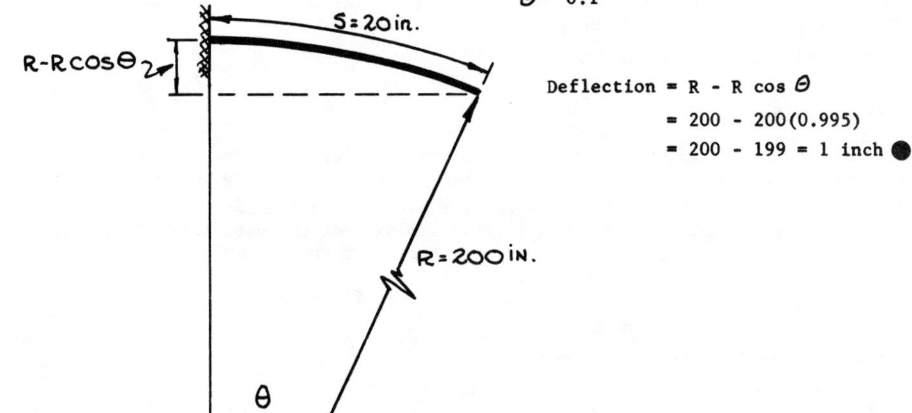

Deflection = $R - R\cos\Theta$

$= 200 - 200(0.995)$

$= 200 - 199 = 1$ inch ●

MECH OF MTLS 53

Fixing both ends of a simply supported beam that has only a concentrated load at midspan will increase the allowable load by:

(a) 25%
(b) 50%
(c) 100%
(d) 200%
(e) no change

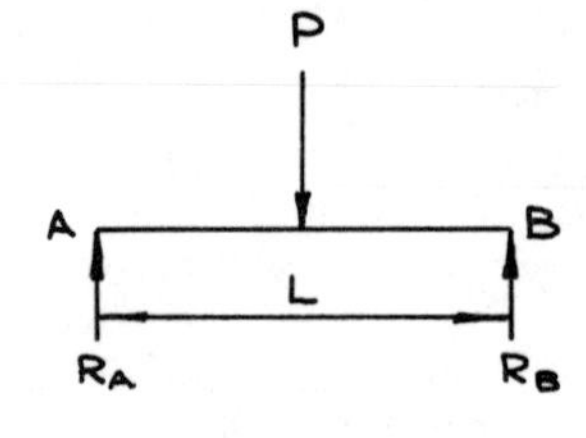

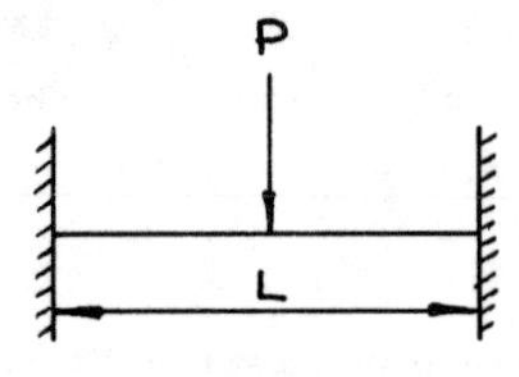

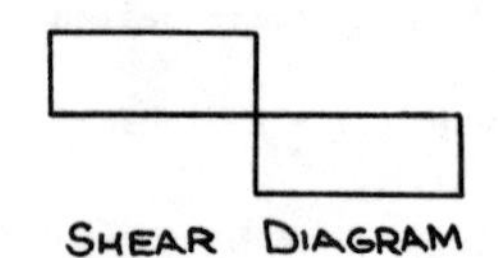

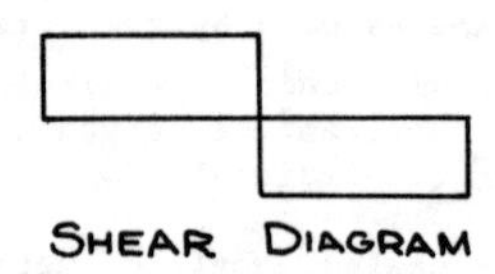

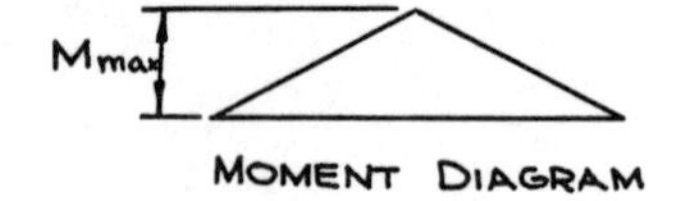

Mmax
Mmax
MOMENT DIAGRAM

$R_A = R_B = \frac{P}{2}$

$M_{max} = \frac{P}{2} \cdot \frac{L}{2} = \frac{PL}{4}$

$R_A = R_B = \frac{P}{2}$

$M_{max} = \frac{P}{2} \cdot \frac{L}{4} = \frac{PL}{8}$

Based on maximum moment, the beam with fixed ends will support twice the applied load of the simple beam.

Answer is (c)

MECH OF MTLS 54

In a long column (slenderness ratio > 160) which of the following has the greatest influence on its tendency to buckle under a compressive load.

(a) The modulus of elasticity of the material
(b) The compressive strength of the material
(c) The radius of gyration of the column
(d) The length of the column
(e) The moment of inertia of the column

Euler's formula for slender columns:

$$P_{crit} = \frac{\pi^2 EI}{L^2}$$

Answer is (d)

MECH OF MTLS 55

The area of the shear diagram of a beam between any two points on the beam is equal to the

(a) Change in shear between the two points
(b) Total shear beyond the two points
(c) Average moment between the two points
(d) Change in moment between the two points
(e) Change in deflection between the two points.

Answer is (d) ●

MECH OF MTLS 56

Poisson's ratio is principally used in

(a) The determination of capability of a material for being shaped
(b) The determination of capacity of a material for plastic deformation without fracture
(c) Stress-strain relationships where stresses are applied in more than one direction
(d) The determination of the modulus of toughness
(e) The determination of the endurance limit

$$\text{Poisson's ratio } \mu = \frac{\text{unit transverse deformation}}{\text{unit axial deformation}}$$

Answer is (c) ●

MECH OF MTLS 57

A railroad track is laid at a temperature of 15°F with gaps of 0.01 feet between the ends of the rails. The rails are 33 feet long. If they are prevented from buckling, what stress will result from a temperature of 110°F?

$\Delta L = \alpha L \Delta T$ where $\alpha = 6.5 \times 10^{-6}$ per °F
$L = 33$ feet
$\Delta T = 110 - 15 = 95°$

$\Delta L = 6.5 \times 10^{-6} \times 33 \times 95 = 0.0204$ feet. This elongation cannot occur and hence there is a stress induced in the rail.

The strain in the rail $= 0.0204 - 0.01 = 0.0104$ ft/33 ft rail
$= 0.00031$ ft/ft (or in/in)

The thermally induced compressive stress:

$$S = \epsilon_t E = 0.00031 \times 30 \times 10^6 = 9300 \text{ psi}$$ ●

MECH OF MTLS 58

The structure shown on the next page is pinned at A and rests on frictionless rollers at B. The structure has equal legs of length L. If both legs are loaded by a triangular load distribution of maximum value w lb/ft, determine:

(a) The reactions at points A and B.
(b) At what point in the structure is the bending moment a maximum?
(c) What is the magnitude of this maximum bending moment?

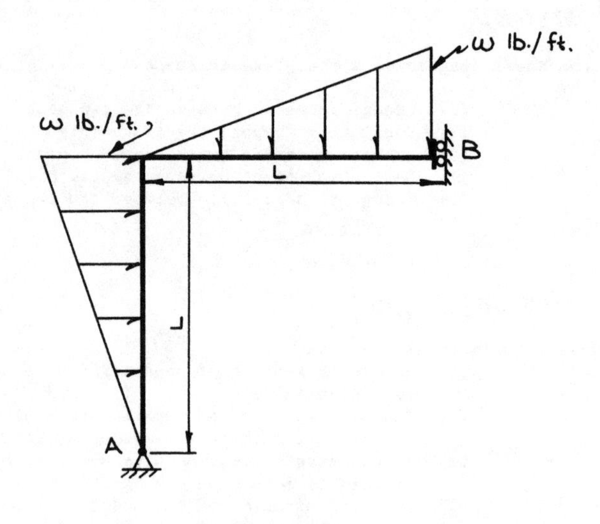

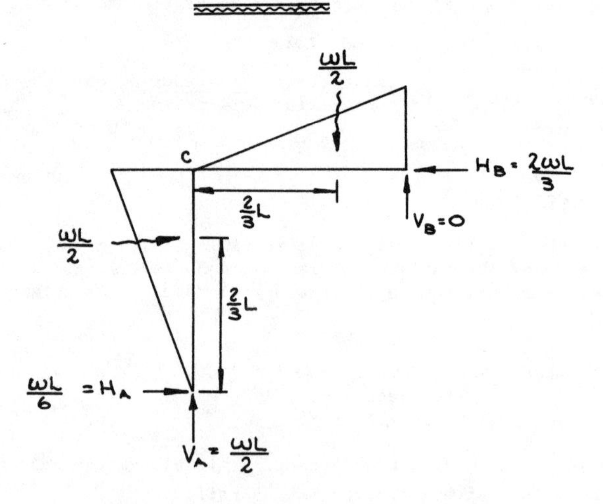

Due to the frictionless rollers at B, $V_B = 0$ ●

$\Sigma M_A = 0$ (+↻) $\quad \frac{wL}{2}(\frac{2}{3}L) + \frac{wL}{2}(\frac{2}{3}L) - H_B(L) = 0$

$$\frac{wL^2}{3} + \frac{wL^2}{3} - LH_B = 0 \qquad H_B = \frac{2wL}{3} \leftarrow$$ ●

$\Sigma F_H = 0$

$$\frac{wL}{2} - \frac{2wL}{3} + H_A = 0 \qquad H_A = \frac{wL}{6} \rightarrow$$ ●

$\Sigma F_V = 0$

$$V_A - \frac{wL}{2} = 0 \qquad V_A = \frac{wL}{2} \uparrow$$ ●

By inspection we see that the maximum bending moment occurs at C ●

$M_{max} = M_C = \frac{wL}{2} \cdot \frac{2}{3}L = \frac{wL^2}{3}$ ●

5

Materials Science

MAT SCI 1

Young's modulus of elasticity for a material can be calculated indirectly from which of the following properties of the material?

(a) Temperature coefficient of expansion and dielectric constant.
(b) Temperature coefficient of expansion and specific heat.
(c) Density and velocity of sound in the material.
(d) Density and interatomic spacing in the material.
(e) Coefficient of conductivity for heat and for electricity.

Speed of longitudinal compressional sound waves in ft/sec is dependent on modulus of elasticity E, lb/in^2. Dimensional comparisons among variables assist in making the proper selection:

(a) α, $(°F^{-1})$ and k, (dimensionless)

(b) α, $(°F^{-1})$ and C_p, $(BTU/lb_m°F)$

(c) ρ, $(slug/ft^3)$ and V, (ft/sec)

(d) ρ, $(slug/ft^3)$ and λ, (ft)

(e) k, $(BTU/(hr\ ft^2)(°F/ft)$ and σ, $(ohm^{-1}\ meter^{-1})$

Since slug $= lb_f sec^2/ft$, manipulation of (c) yields the relationship:

$$E,\ \frac{lb}{in^2} = \rho\left(\frac{lb_f sec^2/ft}{ft^3}\right) V^2\ (ft/sec)^2 \times (1\ ft^2/144\ in^2) = \frac{\rho V^2}{144}$$

If $\rho = lb_m/ft^3$, divide by $g_c = 32.2$ to convert lb_m to slugs:

$$E = \frac{\rho V^2}{(32.2)(144)}$$

Answer is (c) ●

MAT SCI 2

The linear portion of the stress-strain diagram of steel is known as the

(a) modulus of elasticity
(b) plastic range
(c) irreversible range
(d) elastic range
(e) secant modulus

The portion of the diagram where stress is linearly proportional to strain is called the elastic range. Yield strength is usually determined at 0.2% offset, while elastic modulus is calculated from slope of the linear portion of the stress-strain diagram. Non-linear materials such as rubber, plastics, or concrete use a secant modulus of elasticity based on a line from the origin to a predetermined fraction of ultimate tensile or compressive strength.

Tensile Stress - Strain Diagram for ductile steel

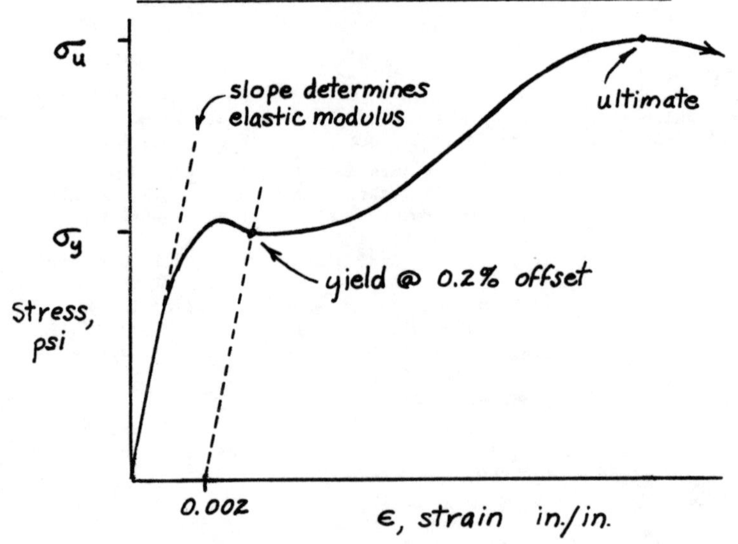

Answer is (d)

MAT SCI 3

"Modulus of Resilience" is

(a) the same as the modulus of elasticity.
(b) a measure of a materials ability to store strain energy.
(c) the reciprocal of the modulus of elasticity.
(d) a measure of the deflection of a member.
(e) a measure of the strain induced in a member due to a change in temperature.

The modulus of resilience is the amount of energy per unit of volume that a given material stores when stressed to the elastic limit.

Answer is (b)

MAT SCI 4

Imperfections within metallic crystal structures may be all of the following, EXCEPT:

(a) lattice vacancies, or extra interstitial atoms.

(b) ion pairs missing in ionic crystals (Shotky imperfections).

(c) displacement of atoms to interstitial sites (Frankel defects).

(d) twinning to form mirror images along a low energy boundary or crystal plane.

(e) linear defects, or slippage dislocations caused by shear.

All are correct except (b). Metals contain atoms at the lattice sites. Salts and other ionic crystals contain ions in a charge-balanced structure at adjacent lattice sites.

The false statement is (b) ●

MAT SCI 5

All of the following statements about strain energy are correct, EXCEPT:

(a) It is caused by generation and movement of dislocations through shear or plastic deformation.

(b) It results from trapped vacancies in the crystal lattice.

(c) It is proportional to length of dislocation, shear modulus, and shortest distance between equivalent lattice sites (points).

(d) It is dependent on energy differences existent between stable regularly spaced sites and the metastable sites around point defects.

(e) It is less for sites at grain boundaries than for internal sites within the crystal structure.

All except (e) are correct. Grain boundaries are high energy areas of mismatch where adjacent crystals come in contact.

The false statement is (e) ●

MAT SCI 6

All of the following statements about diffusion and grain growth are correct, EXCEPT:

(a) Vacancies and interstitial atoms affect diffusion, whose net result is movement of atoms to produce a structure of less strain and of uniform composition.

(b) Diffusion is irreversible and requires an activation energy; its rate increases exponentially with temperature. It follows the diffusion equation where flux equals diffusivity times concentration gradient.

(c) Activation energy for diffusion through structures is inversely proportional to atomic packing factor in the lattice.

(d) Atoms can diffuse both within crystals and across grain (crystal) boundaries.

(e) Grain growth results from diffusion and minimizes total grain boundary area. Large grains grow at the expense of small ones, and grain boundaries move towards their centers of curvature.

All except (c) are correct. When atoms move from one site to another, bonds are broken and reconstituted. During transition an activation energy is required for distortion of the lattice. Small solute atoms, lower melting point solvent, and lower atomic packing factor in the lattice all reduce activation energy. Hence, activation energy for diffusion is directly proportional to packing factor.

The false statement is (c) ●

MAT SCI 7

All of the following statements about slip are correct, EXCEPT:

(a) Slip, or shear along crystal planes, results in an irreversible plastic deformation or permanent set.
(b) It involves only a few atoms at a time in a series of small dislocation movements.
(c) Slip planes lie in the direction of the longest distance between neighboring sites in the crystal lattice.
(d) Ease of slippage is directly related to number of low energy slip planes existent in the lattice structure.
(e) Slip is impeded by solution hardening, wherein large solute atoms serve as anchor points around which slippage does not occur.

Plastic deformation usually occurs by slip. Slip planes lie in the direction of the shortest distance (Burgess vector) between neighboring sites in the crystal lattice.

The false statement is (c) ●

MAT SCI 8

When a metal is cold worked all of the following generally occur EXCEPT:

(a) Recrystallization temperature decreases
(b) Ductility decreases
(c) Grains become equi-axed
(d) Tensile strength increases
(e) Slip or twinning occurs.

Cold working, such as rolling, forging, drawing, or extrusion, reduces cross sectional area at temperatures below the recrystallization temperature. Strain hardening occurs, increasing both yield and ultimate strength. Internal strains and minute cracks are introduced as slip or twinning occur. Ductility, elongation and recrystallization temperature are decreased. A preferred grain orientation is introduced in the direction of elongation, but the grains are not equi-axed.

The false statement is (c) ●

MAT SCI 9

All of the following statements about strain hardening are correct, EXCEPT:

(a) Strain hardening strengthens metals. Resistance to deformation increases with the amount of strain present.
(b) Strain hardening is relieved during softening, annealing above the recrystallization temperature.
(c) Strain hardening is produced by cold working (deformation below the recrystallization temperature).
(d) More strain hardening requires more time-temperature exposure for relief.
(e) Strain hardening is relieved during recrystallization. Recrystallization produces less strained and more ordered structures.

All are correct except (d). Time at or above the recrystallization temperature allows relief of strain hardening. More cold working produces more strain hardening, but this also lowers recrystallization temperature and requires a lesser time-temperature exposure for relief.

The false statement is (d) ●

MAT SCI 10

All of the following processes strengthen metals, EXCEPT:

(a) Annealing above the recrystallization temperature.

(b) Work hardening by mechanical deformation below the recrystallization temperature (cold working).

(c) Precipitation processes, such as age hardening, which produce high strength by formation of submicroscopic phases during low temperature heat treatment.

(d) Heat treatments such as quenching and tempering, for production of a finer microstructure.

(e) Increasing the carbon content of low carbon steels.

All except (a) are correct. Strength and hardness are increased at the expense of ductility and toughness (opposite of brittleness) by developing structures which retard dislocation movement. Annealing increases ductility, but reduces both strength and hardness by relieving strain or work hardening.

The false statement is (a) ●

MAT SCI 11

The valence band model used to explain metallic conduction is based on all of the following statements, EXCEPT:

(a) Each valence band may contain up to 2n electrons/n atoms; each electron lies at a discretely different energy level.

(b) Fermi energy level, E_f, is essentially temperature independent, and is the energy at which 50% of available energy states are occupied.

(c) A conduction band lies at the next higher set of electronic energy levels above those occupied at the ground state.

(d) An energy gap exists between the valence band and the conduction band in semiconductors and insulators.

(e) Conduction occurs when an electron remains in its existent valence band.

All except (e) are correct. Conduction occurs when an electron-hole pair are produced by thermal energy and the electron receives sufficient energy to occupy a vacant and accessible energy state in the higher energy conduction band. In insulators the large energy gap between valence and conduction bands severely reduces the probability of an electron obtaining sufficient energy to jump the gap and to exist in the higher energy conduction band.

The false statement is (e) ●

MAT SCI 12

Intrinsic silicon becomes extrinsically conductive, with electrons as majority carriers, when doped with which of the following?

(a) nothing
(b) antimony
(c) boron
(d) germanium
(e) aluminum

Extrinsic conduction in group IV elements (Silicon or Germanium) is produced by doping with a very small fraction of a ppm of a group III element (Boron or Aluminum) or a group V element (Arsenic or Antimony).

Group III elements have a deficiency of electrons relative to the group IV structure; they produce p-type semiconduction where holes are the majority charge carriers, and any electrons in the valence conduction band are minority carriers.

Group V elements have an excess of electrons, relative to the group IV structure; they produce n-type semiconduction where electrons in the conduction band are majority charge carriers, and holes are minority carriers.

Answer is (b) ●

MAT SCI 13

When the emitter to base of an npn transistor is forward biased, and base to collector is reverse biased, all of the following are correct EXCEPT:

(a) Electrons are majority carriers in the n-emitter and n-collector regions.
(b) Electrons are minority carriers in the p-base region.
(c) Holes are majority carriers in the p-base region.
(d) The emitter is positive with respect to the collector.
(e) Holes are minority carriers in the n-emitter and n-collector regions.

Forward bias increases and reverse bias decreases negative charge on an n-region. Electrons are majority carriers in n-regions and are the minority carriers in p-regions. A forward biased n-type emitter acts as an electron source.

Symbolically, the emitter is shown with an arrowhead in the direction from p- to n-regions. Base is very thin to minimize the recombination of opposite moving electrons and holes, and to reduce their transit times.

In basic circuits the emitter to base is forward biased and base to collector is reverse biased. For an npn transistor (n-emitter, p-base, n-collector), the base must be positive with respect to the emitter, and the emitter must be negative with respect to the collector. For a pnp transistor these bias voltage polarities are reversed.

Based on the above transistor principles, all except (d) are correct.

The false statement is (d) ●

MAT SCI 14

All of the following statements about solid solutions are correct, EXCEPT:

(a) Solid solutions can result when basic structure of the solvent can accomodate solute additions.

(b) In solid solutions larger solute atoms occupy the interstitial space between solvent atoms that are located at the lattice sites.

(c) Solid solutions may result by substitution of one atomic species for another, provided radii and electronic structures are compatible.

(d) Order - disorder transitions that occur at elevated temperature in solid solutions involve changes due to thermal agitation from preferred orientation to random occupancy of lattice sites.

(e) Defect structures exist in solid solutions of ionic compounds because vacancies are required to maintain an overall charge balance when there are differences in oxidation state of solute and solvent ions.

All are correct except (b). Interstitial solid solutions result when very small solute atoms, such as H, C or N, occupy interstitial space between larger solvent atoms located at the lattice sites.

Example: carbon interstitially dissolved in the face centered cubic austenite (γ- iron) structure of steels.

The false statement is (b) ●

MAT SCI 15

All of the following statements about ferromagnetism are correct, EXCEPT:

(a) Magnetic domains are small volumes existent within a single crystal where atomic magnetic moments are unidirectionally aligned.

(b) Domains are randomly oriented when unmagnetized. On magnetization, domains oriented with the external field grow at the expense of unaligned domains.

(c) Impurities, inclusions and strain hardening interfere with change of domain boundaries, and add to the permanency of a magnet.

(d) High magnetic susceptibility of ferromagnetic materials disappears below the Curie temperature.

(e) Above the Curie temperature (770°C, 1418°F for iron) ferromagnetic domains disappear and the metal becomes simply paramagnetic, a consequence of unpaired electrons.

All are correct except (d). Spontaneous magnetism diminishes at elevated temperatures and vanishes above the Curie temperature.

Atomic magnetic moments (Bohr magnetons) arise mainly from unbalanced electron spins in the 3d orbitals.

The false statement is (d) ●

MAT SCI 16

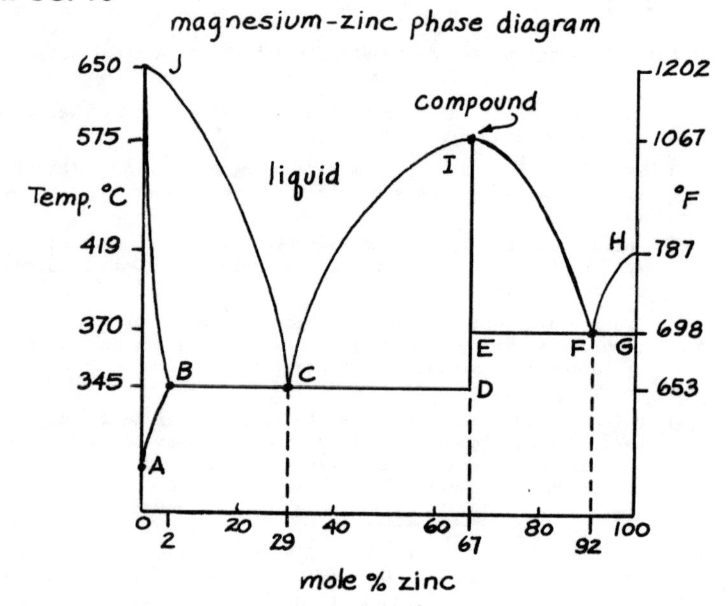

The above phase diagram indicates which of the following intermetallic compounds?

(a) $MgZn_2$

(b) Mg_2Zn

(c) MgZn

(d) Mg_2Zn_3

(e) Mg_3Zn_2

Compound formation splits the phase diagram into separate diagrams to the left and right of the compound composition at 67 mole % Zn and 33 mole % Mg, or $MgZn_2$. The region ABJ represents solubility of the compound in solid magnesium, while C and F are eutectic compositions.

Intermetallic compounds are characterized by nearly perfect structural ordering among unlike neighboring atoms. They have the typical microstructure of a pure metal, an aggregate of crystals of one kind. They are usually brittle and mechanically weak.

Answer is (a) ●

MAT SCI 17

Refer to the magnesium - zinc phase diagram of the previous question. At 575°C a liquid of eutectic composition C (29 mole % Zinc) has sufficient zinc added to raise zinc composition to 40 mole %. When this new liquid is cooled, the first solid phase to separate is:

(a) solid solution containing less than 1% intermetallic compound dissolved in Mg.

(b) solid intermetallic compound

(c) a mixture of solid intermetallic compound and solid eutectic C (29 mole % Zn)

(d) solid eutectic C (29 mole % Zn)

(e) a mixture of solid intermetallic compound and solid eutectic F (92 mole % Zn)

On cooling, the first phase transition occurs on line C-I of the diagram at about 419°C, here much liquid (40 mole % Zn along CI) and a slight amount of solid intermetallic compound (67 mole % Zn along IE) exist. On further cooling to just above 345°C the system contains $\frac{67 - 40}{67 - 29}$ mole fraction of liquid eutectic C (29 mole % Zn) and $\frac{40 - 29}{67 - 29}$ mole fraction of solid intermetallic compound D (67 mole % Zn), computed by tie line lever rule. Further cooling yields solid eutectic C and solid intermetallic compound D at the calculated mole fractions.

Answer is (b) ●

MAT SCI 18

All of the following statements about steels are correct, EXCEPT:

(a) Yield strength of commercially available heat treated alloy steels does not exceed 200,000 psi.

(b) High temperature alloys used in jet engine turbine blades can withstand 2000°F continuously over extended periods.

(c) Abrasion resistance of extra strength steels may be obtained by increasing hardness to 225 - 400 Brinell at the expense of some ductility and toughness.

(d) Intergranular corrosion of chromium - nickel stainless steels is reduced when stabilized by addition of columbium (niobium), titanium or tantalum to preferentially form carbides and prevent chromium depletion and chromium carbide precipitation in grain boundary areas.

(e) Steels used in reinforced concrete construction include ASTM A-15 billet and A-432 high strength billet with tensile yield strengths from 33,000 to 75,000 psi.

Over 100 steels are available with yield strengths above 225,000 psi. Any steel with yield strength above 160,000 psi is considered an ultra-high strength steel.

Examples are: medium carbon quenched and tempered alloy steels
medium alloy air hardening steels
martensitic stainless steels
cold rolled austenitic stainless steels
semi-austenitic stainless steels
high alloy quenched and tempered steels
maraging steels
cold worked plain high carbon steels

The last two types can have ultimate strengths in excess of 300,000 psi.

The false statement is (a) ●

MAT SCI 19

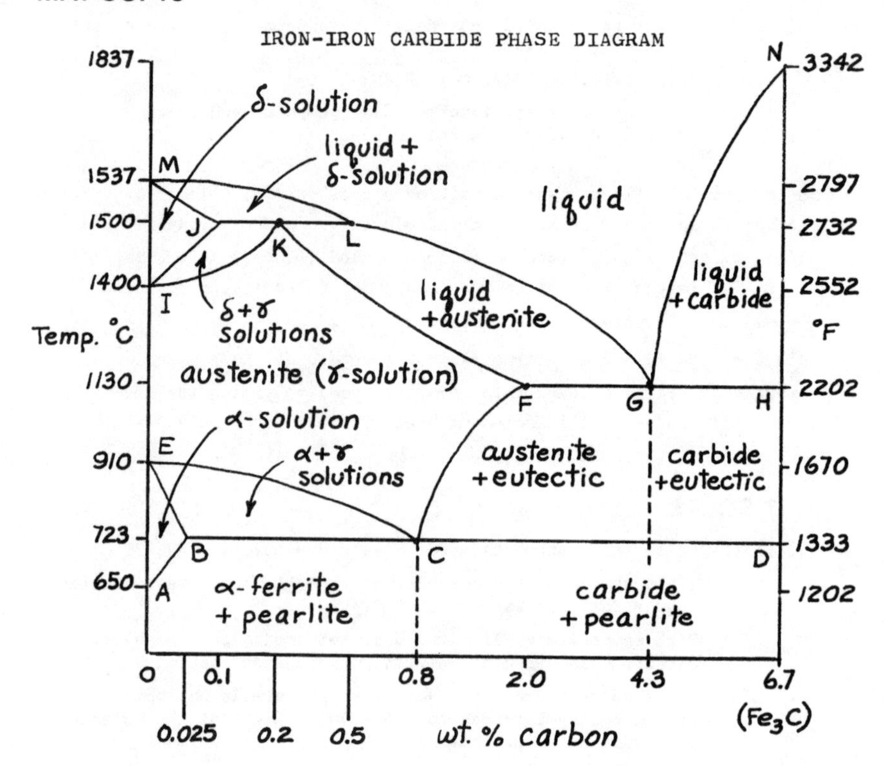

On the usual iron - iron carbide phase diagram above, all of the following are true, EXCEPT:

(a) a eutectic reaction occurs at G, 1130°C.

(b) a peritectic reaction occurs at K, 1500°C.

(c) a eutectoid solid reaction occurs at C, 723°C.

(d) a peritectoid solid reaction occurs at F, 1130°C.

(e) the eutectoid composition is at 0.8 wt.% C.

All are correct except (d). Typical phase reactions involve transition between one phase and two phases. As outlined below, 1, 2 and 3 denote increasing wt.% composition:

(1) eutectic reaction: $\text{Liq}_2 \underset{\text{heat}}{\overset{\text{cool}}{\rightleftharpoons}} \text{Solid}_1 + \text{Solid}_3$

(2) peritectic reaction: $\text{Solid}_1 + \text{Liq}_3 \underset{\text{heat}}{\overset{\text{cool}}{\rightleftharpoons}} \text{Solid}_2$

(3) eutectoid reaction: $\text{Solid}_2 \underset{\text{heat}}{\overset{\text{cool}}{\rightleftharpoons}} \text{Solid}_1 + \text{Solid}_3$

(4) peritectoid reaction: $Solid_1 + Solid_3 \underset{heat}{\overset{cool}{\rightleftharpoons}} Solid_2$

A peritectoid is a rare, reversed eutectoid. No peritectoid exists in this system.

(5) monotectoid reaction: $Liquid_2 \underset{heat}{\overset{cool}{\rightleftharpoons}} Solid_1 + Liquid_3$

A monotectoid is a rare eutectic where the amount of one component in the eutectic composition is nil. No monotectoid exists in this system.

The eutectic at G (4.3 wt.% C) is a mixture of solid austenite (carbide dissolved in γ-iron, composition F) and solid carbide (cementite, composition H).

The peritectic at K is a mixture of solid δ-solution (carbide dissolved in δ -iron, composition J) and liquid solution (composition L).

The eutectoid at C (0.8 wt.% Carbon) yields pearlite, a lamillar structured solid mixture of α solid solution (composition B) and solid carbide (composition D).

The intersection at F indicates composition of the austenite component of eutectic G.

The false statement is (d) ●

MAT SCI 20

Steels can be strengthened by all of the following practices, EXCEPT:

(a) annealing
(b) quenching and tempering
(c) work hardening
(d) grain refinement
(e) age or precipitation hardening

Annealing removes the strength increase caused by work hardening. Quenched and tempered steels rapidly get stronger as carbon content is increased to 0.6%, then more slowly above 0.6%. Alloys can increase strength by increasing hardenability (depth of hardening inward from the quenched surface). Work hardening results from cold drawing, rolling or extrusion performed below the recrystallization temperature. 600,000 psi tensile strength wire is produced by such work hardening of high carbon steel. Combinations of alloying elements can produce fine grained steels that are harder and tougher than course grained steels of the same composition.

The false statement is (a) ●

MAT SCI 21

If 1080 steel (0.80 wt.% C) is annealed by very slow cooling from 1000°C (1832°F) to ambient temperature, its microstructure will consist almost solely of:

(a) austenite
(b) bainite
(c) cementite
(d) martensite
(e) pearlite

1080 steel in equilibrium at 1000°C has the austenite structure. During very slow annealing to ambient, diffusional transformation to pearlite occurs. On rapid quenching to ambient, a shear transformation yields metastable martensite. On tempering martensite at 200 - 400°C, fine grained bainite occurs enroute to course pearlite of the same α-ferrite plus carbide composition. Cementite is iron carbide (Fe_3C) containing 6.67 wt.% carbon.

Answer is (e) ●

MAT SCI 22

All of the following statements about the austenite-martensite-bainite transformations in steel are correct, EXCEPT:

(a) Martensite is fine grained α-ferrite, supersaturated with carbon, in a metastable body centered tetragonal structure. It forms by shear (slippage) during the rapid quench of face centered cubic austenite (γ -ferrite).

(b) Pearlite is a stable and course grained lamillar mixture consisting of body centered cubic α-ferrite plus carbide. It forms by eutectoidal transformation during the slow annealing of austenite. Most alloying elements in steel tend to retard this eutectoidal transformation.

(c) Martensite is strong and hard, but brittle. Tempering toughens it and reduces brittleness.

(d) Tempering of martensite is done by judicious reheating to produce, by diffusion, a fine-grained tough, strong microstructure.

(e) Bainite and tempered martensite have distinctively different microstructures.

Bainite and tempered martensite have very closely related fine-grained microstructures. Pearlite, bainite and tempered martensite all consist of α-ferrite plus carbide. Bainite is usually formed from austenite by partial quench, to avoid pearlite formation, followed by isothermal transformation to fine-grained, tough, strong bainite. Differences between bainite and tempered martensite are subtile.

The false statement is (e) ●

MAT SCI 23

All of the following statements about mechanical failure are true, EXCEPT:

(a) Brittle fracture occurs with little plastic deformation and relatively small energy absorption.

(b) Ductile fracture is characterized by significant amounts of energy absorption and plastic deformation (evidenced by elongation and reduction in cross-sectional area).

(c) Ductile-brittle transition in failure mode occurs at reduced temperatures for most materials, because fracture strength remains constant with temperature while yield strength increases as temperature is reduced. At high temperatures yield strength is least; at low temperatures fracture strength is least.

(d) Fatigue failure due to cyclic stress is frequency dependent.

(e) Creep is time-dependent plastic deformation which accelerates at increased temperature. Stress rupture is the failure resulting from creep.

All are correct except (d). Fatigue failure is not dependent on frequency at ambient temperature. It is very sensitive to surface imperfections from which cracks originate and propagate.

The false statement is (d) ●

MAT SCI 24

All of the following statements about rusting of iron are correct, EXCEPT:

(a) Contact with water, and oxygen are necessary for rusting to occur.

(b) Contact with a more electropositive metal reduces rusting.

(c) Halides aggravate rusting, a process which involves electrochemical oxidation-reduction reactions.

(d) Pitting occurs in oxygen-rich anodic areas, and the rust is deposited nearby.

(e) Strained regions are anodic relative to unstrained regions in the metal.

All except (d) are correct. Pitting occurs in oxygen depleted anodic areas. The localized rust deposit prevents further oxygen diffusion, further aggravating the situation. The cathodic area is not corroded. Cathodic protection makes iron the cathode; if corrosion occurs, it occurs at a sacrificial anode such as a magnesium rod or the zinc galvanizing.

The false statement is (d) ●

MAT SCI 25

Which of the following is not a method of non-destructive testing of steel castings and forgings?

(a) radiography
(b) magnetic particle
(c) ultrasonic
(d) liquid penetrant
(e) chemical analysis

All methods except (e) attempt to detect internal cracks and voids. Chemical analysis is destructive of the sample and does not detect structural flaws in the casting or forging.

The false answer is (e) ●

MAT SCI 26

Compressive strength of fully cured concrete is most directly related to:

(a) sand-gravel ratio
(b) fineness modulus
(c) aggregate gradation
(d) absolute volume of cement
(e) water-cement ratio

Sand and fine material (under 1/4 inch) are "fine aggregate", while gravel and crushed rock (to 3 inches) are "coarse aggregate".
Cement after hydration becomes the adhesive that binds a closely packed aggregate mixture, hence sand-gravel ratio must be optimized to obtain close packing for maximum strength. Fineness modulus is indicative of

the sieve analysis of the fine aggregate, and is used in calculating the ratio of course to fine aggregates used in the mix. Aggregate gradation affects slump or workability, with the larger sizes allowed in thicker sections.

Batch yields are calculated using absolute volumes of cement, water and aggregates. Usual specific gravities are 2.6-2.7 for sand and gravel, and 3.1 for cement.

Water-cement ratio is the most important factor affecting strength of the concrete; less water yields a higher strength, though less workable (lower slump) concrete.

Answer is (e) ●

MAT SCI 27

According to the ACI code, the modular ratio, n, of structural concrete with a 28-day ultimate compressive strength, f'_c, of 3000 is nearest to:

(a) 7
(b) 8
(c) 9
(d) 10
(e) 11

The modular ratio n is the ratio of elastic moduli of steel to concrete, E_s/E_c.

E_s = 29,000,000 psi, while E_c is calculated from concrete weight w and and strength f'_c using: $E_c = w^{1.5} \times 33(f'_c)^{0.5}$

The strength f'_c ranges from 2000 - 5000 psi, while concrete weight W varies with aggregate density. Lightweight aggregates (dry weights under 70 lb/ft^3) yield structural lightweight concrete of 90 - 115 lb/ft^3, while sand and gravel (about 100 lb/ft^3) yield structural concrete at about 145 lb/ft^3.

$$E_c = 145^{1.5}(33)(3000)^{0.5} = 3.2 \times 10^6 \text{ psi}$$

$$n = \frac{E_s}{E_c} = \frac{29 \times 10^6}{3.2 \times 10^6} = 9$$

Answer is (c) ●

MAT SCI 28

All of the following statements about air entrained concrete are correct, EXCEPT:

(a) Air entrainment is recommended when concrete is exposed to severe frost action.

(b) With air entrainment, the quantity of water to produce a given consistency (slump) is reduced.

(c) With air entrainment, the quantity of water to produce a specified 28 day compressive strength is reduced.

(d) Air entrainment reduces friction between solids during mixing.

(e) Air entrainment reduces resistance to the freeze-thaw that occurs when salt is used to melt ice or snow.

The amount of entrained air varies inversely with maximum aggregate size, ranging from 0.2 to 3% without, and 3 to 8% with an added air entrainment agent. Voids reduce strength, and for a given water-cement ratio, air entrainment can reduce strength as much as 20%. Air entrainment does increase resistance to freeze-thaw conditions.

The false statement is (e) ●

MAT SCI 29

In the design of a reinforced concrete structure, tensile strength of the concrete is normally:

(a) Assumed to be 1/10 of the 28-day compressive strength.

(b) Determined by beam tests.

(c) Neglected.

(d) Assumed to be 200 psi.

(e) Determined by split cylinder tests.

Hairline cracks can reduce tensile strength to zero. Reinforced concrete design by conventional cracked section analysis assumes no tensile strength in the concrete and takes all tensile forces in the steel. (Split cylinder tests do yield design data that are used to guard against diagonal tension failure resulting from shear stresses.) Prestressed concrete is reinforced concrete with imposed compressive stresses such that tensile stresses from service loading are opposed and the concrete remains in compression.

Answer is (c) ●

MAT SCI 30

All of the following groups of plastics are thermoplastic, EXCEPT:

(a) Polyvinylchloride (PVC) and polyvinyl acetate

(b) Polyethylene, polypropylene, and polystyrene

(c) Tetrafluoroethylene (Teflon) and other fluorocarbons

(d) Phenolics, melamine and epoxy

(e) Acrylic (Lucite) and polyamide (nylon)

All are thermoplastic except group (d), which are thermosetting. Thermoplastic materials are completely polymerized, but soften for molding at elevated temperatures.

Thermosetting materials complete their polymerization during heating to give hard plastics.

All have different maximum service temperatures, above which decomposition begins. Most plastics become brittle at low temperatures; plasticizers are added to reduce low-temperature brittleness and improve ductility.

The false answer is (d) ●

6

Fluid Mechanics

FLUIDS 1

Kinematic viscosity may be expressed as:

(a) $\frac{ft^2}{sec}$

(b) $\frac{sec^2}{ft}$

(c) $\frac{slug\ sec^2}{ft}$

(d) $\frac{slugs}{sec}$

(e) $\frac{slug\ ft}{sec^2}$

Kinematic viscosity $\nu = \frac{\text{absolute viscosity}}{\text{density}} = \frac{\mu}{\rho}$

Units of viscosity: $lb_f\text{-sec/ft}^2$ or slug/ft-sec

Units of density: $lb_f\text{-sec}^2/ft^4$ or $slugs/ft^3$

In either units the dimensions of kinematic viscosity would be:

$$\frac{\frac{slug}{ft\text{-}sec}}{\frac{slug}{ft^3}} = \frac{ft^2}{sec} \qquad \frac{lb_f\text{-sec/ft}^2}{lb_f\text{-sec}^2/ft^4} = \frac{ft^2}{sec}$$

$slug = lb_f sec^2/ft = lb_m/g_c$ Answer is (a)

FLUIDS 2

Absolute viscosity of a fluid varies with pressure and temperature and is defined as a function of

(a) Density and angular deformation rate.
(b) Density, shear stress and angular deformation rate.
(c) Density and shear stress.
(d) Shear stress and angular deformation rate.
(e) Angular deformation rate only.

Absolute viscosity is proportional to shear stress divided by angular deformation rate. Density is not involved in the definition.

rate of angular deformation $\cong \frac{dU}{dy}$

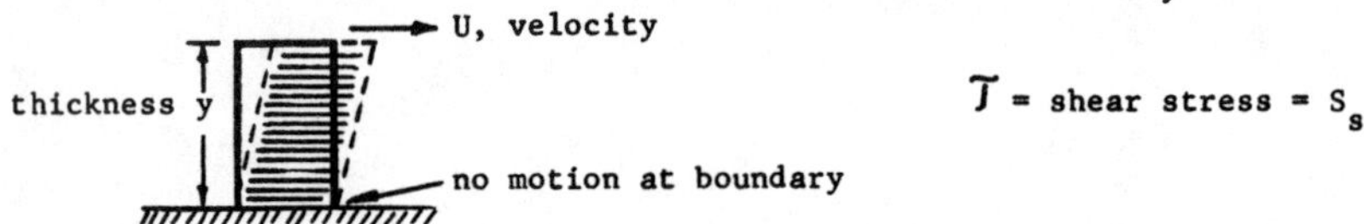

τ = shear stress = S_s

$$\text{absolute } \mu = \frac{S_s \text{ or } \tau}{\frac{dU}{dy}} = \frac{lb_f/ft^2}{\frac{ft/sec}{ft}} = \frac{lb_f\ sec}{ft^2} = \frac{slug}{ft\ sec} \quad \text{in english units}$$

$$1\ \frac{slug}{ft\ sec} = 479.\ poise, \qquad poise = \frac{dyne\ sec}{cm^2} = \frac{gm}{cm\ sec}$$

For Newtonian fluids the relation between shear stress and angular deformation rate is linear; for non-Newtonian and thixotropic materials it is non-linear.

Answer is (d) ●

FLUIDS 3

An open chamber rests on the ocean floor in 160. feet of sea water (sp. gr. = 1.03). What air pressure in psig must be maintained inside to exclude water?

(a) 45.2
(b) 60.9
(c) 71.4
(d) 93.2
(e) 140.0

Internal pressure must equal external pressure.

External P = 160(62.4)(1.03) = 10,280 lb_f/ft^2

$$P = \frac{10,280}{144} = 71.4 \text{ psig}$$ ●

Answer is (c)

FLUIDS 4

What is static gage pressure in psf in the air chamber of the container below?

(a) -200. lb_f/ft^2
(b) -75. lb_f/ft^2
(c) 0
(d) +75. lb_f/ft^2
(e) +200. lb_f/ft^2

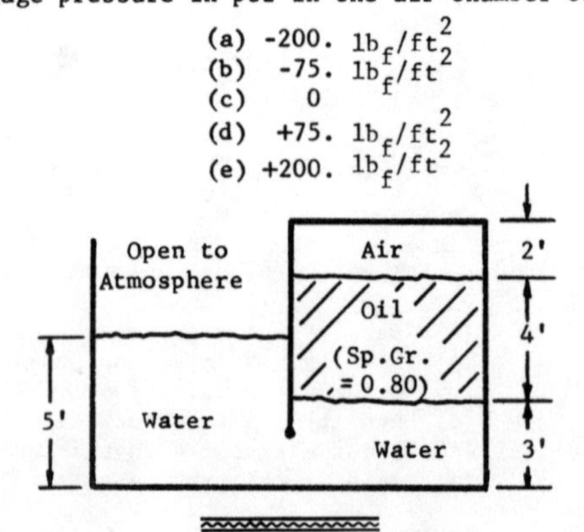

Since the situation is static, gage pressure at the base is 5' water. In the air chamber it is 5' water less 3' water less 4' oil.

P = 2' water - 4' oil = 2(62.4) - 4(62.4)(0.80)
= 124.8 - 199.7 = -74.9 lb_f/ft^2 ●

Answer is (b)

FLUIDS 5

Pressure in lb_f/ft^2 at a depth of 300 feet in fresh water is nearest to:

(a) 5,360 psf
(b) 12,900 psf
(c) 18,700 psf
(d) 32,400 psf
(e) 56,500 psf

Pressure = weight of fluid above = 300(62.4) = 18,720 lb_f/ft^2 ●

Answer is (c)

FLUIDS 6

What head in feet of air, at ambient conditions of 14.7 psia and 68°F, is equivalent to 2 psia?

(a) 146.
(b) 395.
(c) 771.
(d) 1560.
(e) 3840.

Density of air is calculated from ideal gas law using $R = 53.34 \frac{ft\ lb_f}{lb_m\ {}^\circ R}$

$$\rho = \frac{P}{RT} = \frac{(14.7\ lb/in^2)(144\ in^2/ft^2)}{53.34\ ft\ lb_f/(lb_m\ {}^\circ R)(528^\circ R)} = 0.0752 \frac{lb_m}{ft^3}$$

2 psia = 2(144) = 288 lb_f/ft^2

$$h = \frac{288}{0.0752} = 3830 \text{ feet} ●$$

note that specific weight of air
$\gamma = \rho(g/g_c) = 0.0752\ lb_f/ft^3$

Alternately: 1000 feet of air at ambient conditions is approximately equivalent to 0.5 psia.

Answer is (e)

FLUIDS 7

With a normal barometric pressure at sea level, atmospheric pressure at an elevation of 4000 feet is nearest to:

(a) 26" Hg.
(b) 27" Hg.
(c) 28" Hg.
(d) 29" Hg.
(e) 30" Hg.

Normal barometric pressure at sea level is 29.92" Hg.
Near sea level this is reduced by about 1" Hg or 0.5 psia per 1000 feet.

The rate of pressure reduction diminishes with increased altitude since air density decreases with altitude.

Answer is (a) ●

FLUIDS 8

The funnel shown is full of water. The volume of the upper part is 5.83 cu. ft. and of the lower part is 2.50 cu. ft.
The force tending to push the plug out is:

(a) 250 lbs.
(b) 312 lbs.
(c) 380 lbs.
(d) 442 lbs.
(e) 500 lbs.

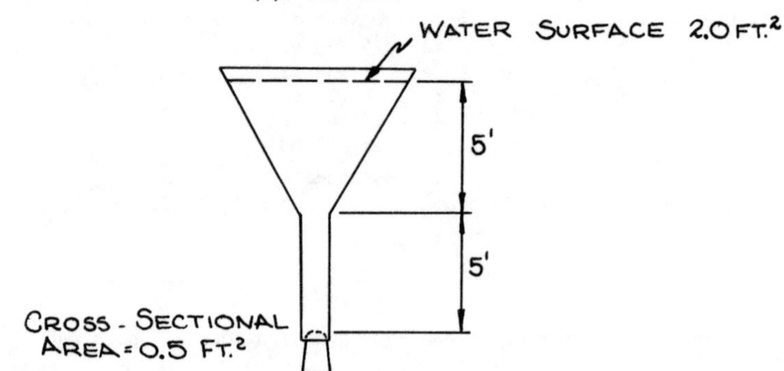

$$\text{Force} = \gamma hA = 62.4\ \frac{lb_f}{ft^3} \times 10\ ft \times 0.5\ ft^2 = 312\ lb_f$$

Answer is (b)

FLUIDS 9

In the sketch shown below, a barrel whose empty weight is 56 lbs. contains an unknown quantity of water in which a container holding 200 lbs. of oil (S.G. = 0.95) floats, and a beaker containing 10 lbs. of mercury (S.G. = 13.0) which floats in the oil. Also, an iron rod is immersed in the water so that it displaces 1.0 cu. ft. of the liquid.

If the beaker, the oil container, and the barrel are 6 inches, 24 inches, and 34 inches in diameter, respectively, and the total weight indicated by the scale is 840 lbs, determine the height h of the water in the barrel.

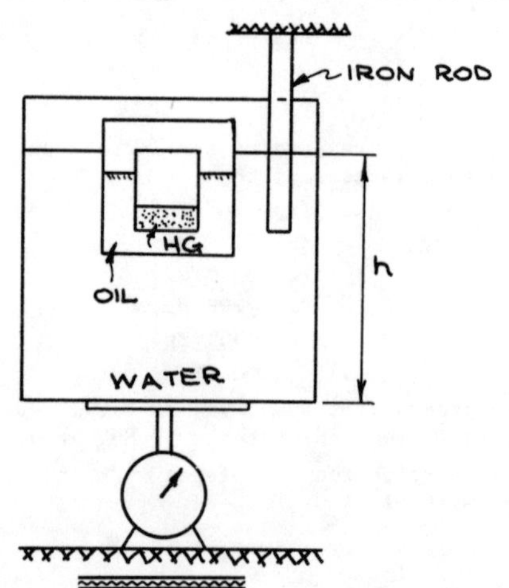

Equivalent volume of water in barrel

$$= \frac{840 \text{ lb} - 56 \text{ lb}}{62.4 \text{ lb/cu ft}} = \frac{784}{62.4} = 12.55 \text{ cu ft}$$

Determine height h for an equivalent volume of 12.55 cu ft

$$V = \pi r^2 h \qquad h = \frac{V}{\pi r^2} = \frac{12.55}{\pi\left(\frac{17}{12}\right)^2} = \frac{12.55}{6.31} = 1.99 \text{ feet}$$ ●

FLUIDS 10

An open topped cylindrical water tank has a horizontal circular base 10. feet in diameter. When filled to a height of 8 feet, the force in lbs. exerted on its base is nearest to:

(a) 3,900.
(b) 7,800.
(c) 10,000.
(d) 26,000.
(e) 39,000.

Pressure at tank base = 8' water = 8(62.4) = 499 lb_f/ft^2

Area of tank base = $\frac{\pi}{4}(10)^2 = 25\pi = 78.5 \text{ ft}^2$

Force on tank base = PA = 499(78.5) = 39,200 lb_f. ●

Answer is (e)

FLUIDS 11

A cubical tank with 4 foot sides is filled with water. The force in lbs. developed on one of the vertical sides is nearest to:

(a) 500.
(b) 1000.
(c) 2000.
(d) 3000.
(e) 4000.

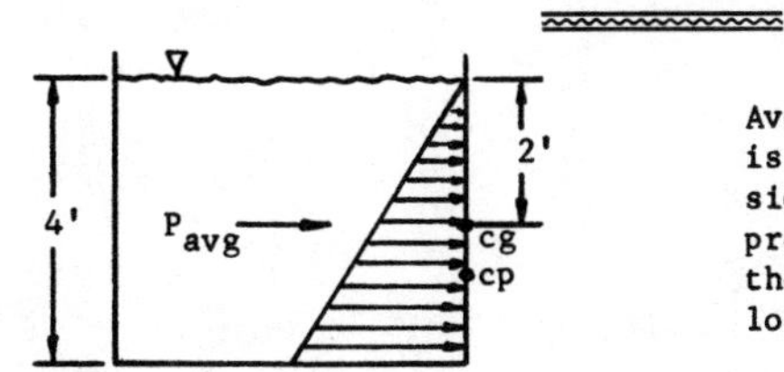

Average pressure exerted on one side is pressure existent at c.g. of the side times area of the side. Since pressure varies linearly with depth, the c.p. (center of pressure) is located below the c.g.

<u>For other problems</u>: $y_{cp} = y_{cg} + \frac{I}{y_{cg}A}$

where y_{cg} = either vertical or slant distance in feet from free fluid surface to c.g. of area considered.

y_{cp} = distance to c.p., same vertical or slant distance basis, ft.

A = area considered, ft^2

I = area moment of inertia about centroid of area considered, ft^4

The c.p. is the location where a single concentrated force can be used to replace the distributed pressure forces for moment purposes. The pressure existent at the area centroid times area equals force exerted normal to the area acting through the c.p.

For this problem: $F = (P_{avg})(A) = 2(62.4)(4 \times 4) = 1997\ lb_f$ ●

Answer is (c)

FLUIDS 12

Each connection between a 90 degree elbow and the 6" diameter pipeline to which it is connected must resist what net tensile force in lbs. under static, no-flow conditions if the line is pressurized to 100 psig?

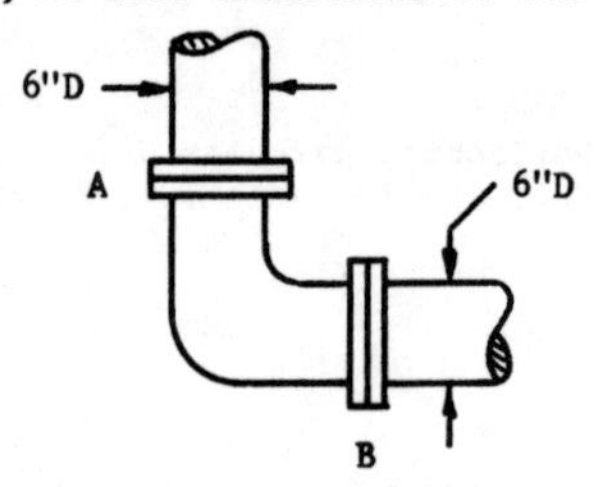

(a) 710.
(b) 830.
(c) 2000.
(d) 2800.
(e) 4000.

Static pressure force at A: $F = (100\ \frac{lb_f}{in^2})\left[\frac{\pi}{4}(6)^2\right] = 900\pi\ lb_f \downarrow$

Static pressure force at B: $F = 900\pi\ lb_f \leftarrow$

Connectors at each joint are under tensile force of $2827\ lb_f$ ●

Answer is (d)

FLUIDS 13

A conical reducing section connects an existent 4" diameter pipeline with a new 2"D line. At 100 psig static pressure under no-flow conditions, what tensile force in lbs. is exerted on the connectors at joint A ?

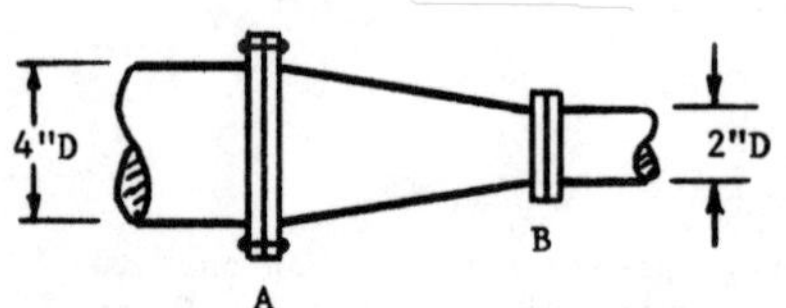

(a) 320.
(b) 470.
(c) 630.
(d) 940.
(e) 1260.

Static force at A $= (100\ \frac{lb}{in^2})\left[\frac{\pi}{4}(4)^2\right] = 400\pi = 1257\ lb_f$ tension on bolts at A ●

Static force at B $= (100\ \frac{lb}{in^2})\left[\frac{\pi}{4}(2)^2\right] = 100\pi = 314\ lb_f$ tension on bolts at B

End restraint by the pipes opposes a net force of $1257-314=943\ lb_f$ to the right on the reducing section.

Answer is (e)

FLUIDS 14

A circular access port 2 feet in diameter seals an environmental test chamber that is pressurized to 15 psi above external pressure. What force in lbs. does the port exert upon its retaining structure?

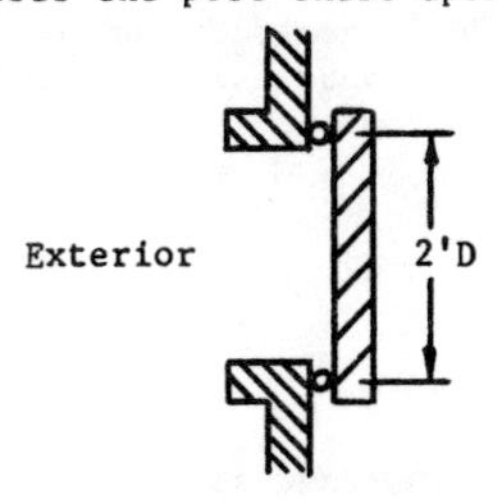

(a) 1700.
(b) 2300.
(c) 3700.
(d) 5400.
(e) 6800.

Area of port $= \frac{\pi}{4} D^2 = \pi \, ft^2$

Pressure $= 15 \frac{lb}{in^2} \times \frac{144 \ in^2}{ft^2} = 2160 \ lb_f/ft^2$

$F = PA = \pi(2160) = 6786 \ lb_f$ ●

Answer is (e)

FLUIDS 15

A vertical sliding gate 20' wide and 30' high is submerged in 40' of water, has a coefficient of friction μ between its guides and its edges of 0.20, and weighs 6 tons. What vertical force is required to just lift this gate? (Assume friction is due to the normal force of the water pressure on the gate and neglect any buoyant force of the water.)

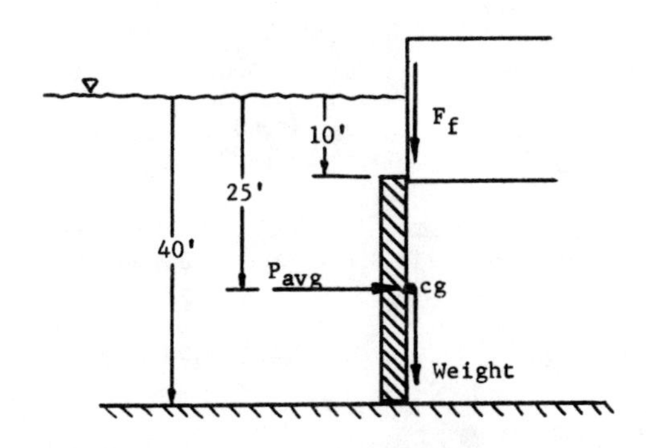

(1) Horizontal force against gate = (average pressure = pressure at area centroid) x area

c.g. is 25' below surface. $P = 25(62.4) = 1560 \ lb_f/ft^2$

$F = PA = 1560(20)(30) = 936{,}000. \ lb_f$

(2) Vertical frictional force $F_f = \mu N$ where μ = coeff. of friction = 0.2

N = force normal or perpendicular to gate = 936,000. lb_f

$F_f = (0.2)(936{,}000) = 187{,}200 \ lb_f$

(3) Total force to start gate upward = its weight + 187,200 lb_f.

= 12,000 + 187,200 = 199,200 lb ●

FLUIDS 16

Commercial propane stored in a spherical steel tank at 100°F generates a gage pressure of 160 psi. If the tank is 4 feet in diameter and has walls 0.25 inch thick, what maximum tensile stress in psi is developed in the steel?

(a) 1,700.
(b) 2,900.
(c) 5,400.
(d) 7,700.
(e) 13,000.

A spherical tank, with no cylindrical section, is twice as strong as a cylindrical tank with hemispherical ends.

$$S_T = \frac{1}{2}\frac{Pr}{t}$$

where P = internal-external pressure difference
r = radius
t = wall thickness
S_T = tensile stress in the wall

all in consistent units.

$$S_T = \frac{1}{2}(160\ \frac{lb_f}{in^2})(\frac{24\ in}{0.25\ in}) = 7,700\ psi$$ ●

Answer is (d)

FLUIDS 17

The steel cylindrical butane storage tank shown in the drawing has hemispherically domed ends. If butane has a vapor pressure of 45. psia at 100°F, and the tank walls are 0.25 inches thick, the maximum tensile stress in psi developed in the tank is nearest to:

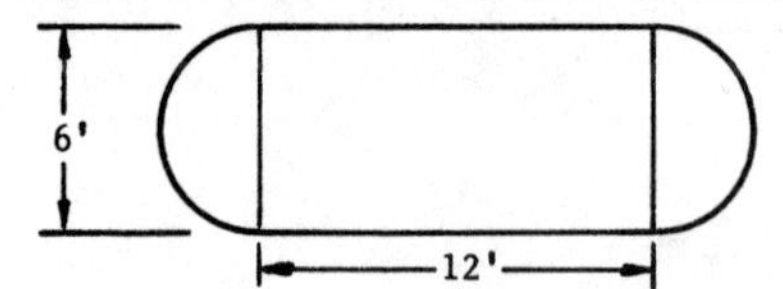

(a) 4,400.
(b) 7,300.
(c) 9,700.
(d) 12,000.
(e) 17,000.

Mode of failure will be a longitudinal rip in the cylindrical section.

Maximum stress will be tensile hoop stress given by $S_T = \frac{Pr}{t}$

P is the internal-external pressure differential = 45 - 14.7 = 30.3 psig

$$S_T = 30.3\ psi(\frac{36}{0.25}) = 4,360\ psi$$ ●

Answer is (a)

FLUIDS 18

A gas bubble rising from the ocean floor is 1 inch in diameter at a depth of 50 feet. Given that sp. gr. of seawater is 1.03, the buoyant force in lbs being exerted on the bubble at this instant is nearest to:

(a) 0.014
(b) 0.020
(c) 0.076
(d) 0.14
(e) 0.20

Volume of bubble = volume of displaced seawater

$$= \frac{4}{3}\pi r^3 = \frac{4}{3}\pi\left(\frac{1}{2}\right)^3 = \frac{\pi}{6} = 0.524\ in^3$$

density of seawater = $62.4(1.03)\frac{lb_m}{ft^3} \times \frac{ft^3}{1728\ in^3} = 0.0372\ \frac{lb_m}{in^3}$. Specific weight of seawater is $0.0372\ lb_f/\ in^3$.

Ignore sp. wt. of bubble calculated at water temperature and an absolute pressure of atmospheric + $50(1.03)(62.4)\ lb_f/ft^2$.

Buoyant force = $0.524\ in^3(0.0372\ \frac{lb_f}{in^3}) = 0.0195\ lb_f$ ●

Answer is (b)

FLUIDS 19

Draft, in inches of water differential pressure, is generated at the base of a 100 foot stack filled with 500°F gases (assume same molecular weight as air) due to differential specific weight.

(Under ambient conditions of 14.7 psia and 68°F, air sp. wt. is $0.075\ lb_f/ft^3$.)

The draft is nearest to:

(a) 0.30" water
(b) 0.65" water
(c) 1.00" water
(d) 1.50" water
(e) 3.00" water

Sp. wt. of stack gases = $0.075\ \frac{528°R}{960°R} = 0.0413\ lb_f/ft^3$

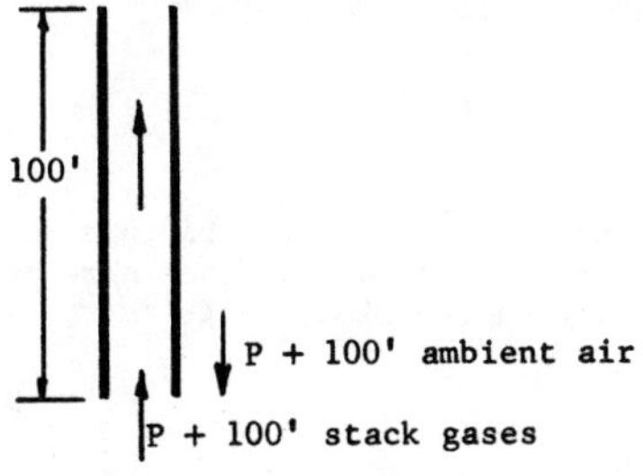

$\Delta P = 100(0.075 - 0.041) = 3.4\ lb_f/ft^2$

1" water = $\frac{62.4}{12} = 5.2\ lb_f/ft^2$

$\therefore \Delta P = \frac{3.4}{5.2} = 0.65"$ water ●

Answer is (b)

FLUIDS 20

Ice in an iceberg has a sp. gr. of 0.922. When floating in seawater (sp. gr. 1.03), its exposed volume % is nearest to:

(a) 5.6
(b) 7.4
(c) 8.9
(d) 10.5
(e) 11.8

Buoyant force equals weight of fluid displaced. At equilibrium or floating, weight downward equals buoyancy.

Let V_1 = total volume of iceberg in ft^3

Its weight is $V_1(62.4)(0.922) = 57.5(V_1)\ lb_f$

Let V_2 = immersed volume of iceberg = volume of seawater displaced.

Weight of seawater displaced = $V_2(62.4)(1.03) = 64.2(V_2)\ lb_f$

$$\frac{V_2}{V_1} = \frac{57.5}{64.2} = 0.895 = \text{volume fraction of iceberg immersed}$$

$\therefore$ volume fraction exposed = 1. - 0.895 = 0.105 = 10.5% ●

Answer is (d)

FLUIDS 21

A cylinder of cork is floating upright in a container partially filled with water. A vacuum is applied to the container such that the air within the vessel is partially removed. The cork will

(a) rise somewhat in the water
(b) sink somewhat in the water
(c) remain stationary
(d) turn over on its side
(e) sink to the bottom of the container

Archimedes' Principle applies equally well to gases. Thus a body located in any fluid, whether liquid or gaseous, is buoyed up by a force equal to the weight of the fluid displaced. A balloon filled with a lighter than air gas readily demonstrates the buoyant force.

Thus the weight of the cork is equal to the weight of water displaced plus the weight of air displaced. When the air within the vessel is removed, the cork is no longer provided a buoyant force equal to the weight of air displaced. For equilibrium, the cork will sink somewhat in the water. ●

Answer is (b)

FLUIDS 22

A floating cylinder 8 cm in diameter and weighing 950 grams is placed in a cylindrical container 20 cm in diameter partially full of water. The increase in the depth of water in the container due to placing the float in it is:

(a) 10 cm
(b) 5 cm
(c) 3 cm
(d) 2 cm
(e) 1 cm

The 950 gram cylinder will displace 950 grams of water. Since 1 cc of water weighs 1 gram, the cylinder will displace 950 cc of water. The change in total volume beneath the water surface dV equals the area of the cylindrical container A times the change in water level dh, or $dV = A\,dh$.

The depth of the water will increase

$$dh = \frac{dV}{A} = \frac{950\ cm^3}{\frac{\pi}{4}(20)^2} = 3.02\ cm$$ ●

Answer is (c)

FLUIDS 23

A block of wood floats in water with 6 inches projecting above the water surface. If the same block were placed in alcohol of specific gravity 0.82, the block would project 4 inches above the surface of the alcohol. Determine the specific gravity of the wood block.

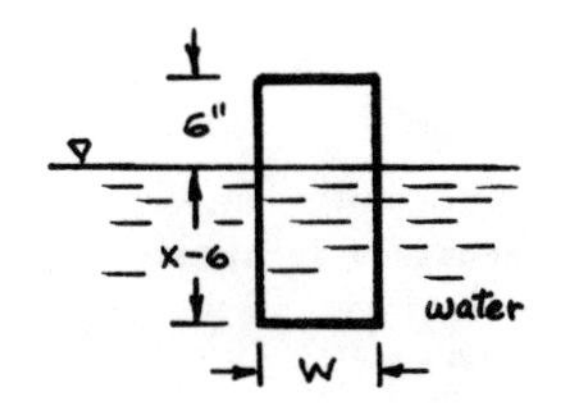

Let x = height of wood block
W = width of wood block
L = length of wood block
γ = specific weight of water, lb_f/ft^3

The weight of the block is equal to the weight of the liquid displaced.

Weight of the block in water $= (x - 6)WL\gamma\ 1.0$

Weight of the block in alcohol $= (x - 4)WL\gamma\ 0.82$

Since the weight of the block is constant:

$$(x - 6)WL\gamma = 0.82(x - 4)WL\gamma$$

$$x - 6 = 0.82x - 3.28 \qquad x = \frac{2.72}{0.18} = 15.1 \text{ inches}$$

The Specific Gravity of the wood block:

$$= \frac{\text{Volume of water displaced}}{\text{total volume}} = \frac{(x - 6)WL}{xWL} = \frac{15.1 - 6}{15.1} = 0.603$$ ●

FLUIDS 24

Average velocity in a full pipe of incompressible fluid at Section 1 of the drawing is 10 ft/sec. After passing through a conical section that reduces the stream's cross-sectional area at Section 2 to one fourth of its previous value, the velocity at Section 2, in ft/sec, is:

(a) 2.5
(b) 5.
(c) 10.
(d) 40.
(e) 80.

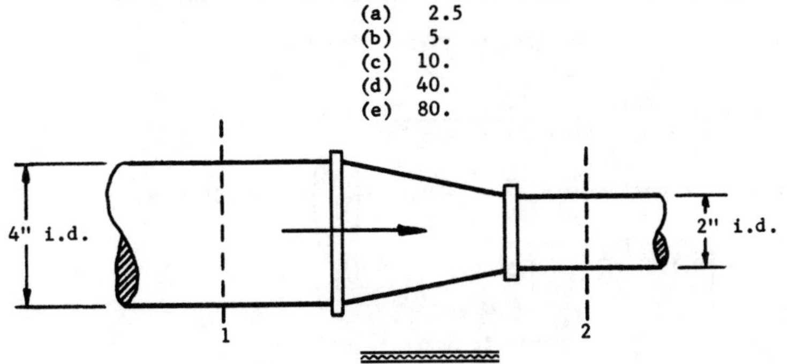

Continuity requires that:

$$q = A_1v_1 = A_2v_2 = A_1(10) = \frac{A_1}{4}\ v_2 \qquad v_2 = 40.\ \text{ft/sec}$$ ●

Answer is (d)

FLUIDS 25

Refer to the drawing of the previous question.
If the static pressure at A is 100 psig and the 4" diameter pipe is full of water in turbulent flow at an average velocity of 30. ft/sec at A, the mass flow rate in lb/sec at B is nearest to:

(a) 3.7
(b) 48.
(c) 160.
(d) 270.
(e) 520.

Basis of calculation: 1 second

Continuity requires that mass flow rate be the same at all sections in steady flow.

Calculate $\dot{m}$ at Section A where velocity is given, using $\dot{m} = \rho Av$

cross-sectional area $A = \frac{\pi}{4}\left(\frac{4}{12}\right)^2 = 0.087\ ft^2$

$$\dot{m} = \left(62.4\ \frac{lb_m}{ft^3}\right)(0.087\ ft^2)\left(30\ \frac{ft}{sec}\right) = 163\ lb_m/sec$$ ●

Answer is nearest (c)

FLUIDS 26

Air flows in a long length of 1" diameter pipe. At one end the pressure is 30 psia, the temperature is 300°F and the velocity is 30 ft/sec.

At the other end the pressure has been reduced by friction and heat loss to 20 psia. The mass flow rate in lb/sec at any section along the pipe is nearest to:

(a) 0.02
(b) 0.11
(c) 0.37
(d) 1.5
(e) 3.9

Mass flow rate $\dot{m} = q\rho = \rho Av$ Basis of calculation: 1 second

Density of air at 30 psia and 300°F (760°R) is obtained from ideal gas law:

$Pv = RT = \frac{P}{\rho}$ Use $R = 53.34\ ft\ lb_f/(lb_m{}^\circ R)$ for air

$$\rho = \frac{P}{RT} = 30(144)\frac{lb_f}{ft^2}\ \frac{lb_m\ {}^\circ R}{53.34\ ft\ lb_f(760^\circ R)} = 0.107\ \frac{lb_m}{ft^3}$$

$$A = \text{cross-sectional area} = \frac{\pi}{4}D^2 = 0.785\left(\frac{1}{12}\right)^2 = 0.00545\ ft^2$$

$$\dot{m} = \rho Av = \left(0.107\ \frac{lb_m}{ft^3}\right)\left(0.00545\ ft^2\right)\left(30\ \frac{ft}{sec}\right) = 0.0175\ \frac{lb_m}{sec}$$ ●

Answer is nearest (a)

FLUIDS 27

Water flows through a long $\frac{1''}{2}$ i.d. hose at 3 gallons per minute.

Water velocity in ft/sec is nearest to:

(a) 1.
(b) 5.
(c) 10.
(d) 20.
(e) 50.

Basis of calculation: 1 second

$$q = 3\ \frac{gal}{min} \times \frac{min}{60\ sec} \times \frac{ft^3}{7.48\ gal} = 0.00668\ cfs$$

$$A = \text{cross-sectional area} = \frac{\pi}{4} D^2 = 0.785 \left[\frac{0.5}{12}\right]^2 = 0.00136\ ft^2$$

$$q = Av \qquad v = \frac{0.00668}{0.00136} = 4.9\ ft/sec$$ ●

Answer is nearest (b)

FLUIDS 28

An incompressible fluid gasoline (ρ = 50 lb/ft^3) enters and leaves a pump system with the following energy in ft lb_f per lb_m of fluid:

	Entering	Leaving
Potential energy Z ft above datum	5	15
$\frac{v^2}{2g_c}$ Kinetic energy	5	10
Pv Flow energy	30	150
Total energy	40	175

The pressure increase in psi between entering and leaving streams is nearest to:

(a) 42.
(b) 84.
(c) 108.
(d) 120.
(e) 135.

Pressure (flow) energy change is 150 - 30 = 120 ft lb_f/lb_m

$$\Delta(Pv) = \frac{\Delta P}{\rho} = 120 \qquad \Delta P = 120 \frac{ft\ lb_f}{lb_m}\ 50\ \frac{lb_m}{ft^3} = 6000\ \frac{lb_f}{ft^2}$$

$$\Delta P = 6000\ \frac{lb_f}{ft^2} \times \frac{ft^2}{144\ in^2} = 41.7\ psi$$ ●

Answer is (a)

FLUIDS 29

Use the data of the previous question. If the volume flow rate of the gasoline ($\rho = 50\ lb/ft^3$) is 15 gallons per minute, theoretical pumping power in HP is nearest to:

(a) 0.4
(b) 0.8
(c) 1.2
(d) 3.4
(e) 6.9

Basis of calculation: 1 minute

$$\text{mass flow rate} = 15\ \frac{gal}{min} \times \frac{ft^3}{7.48\ gal} \times \frac{50\ lb_m}{ft^3} = 100.\ \frac{lb_m}{min}$$

Ignoring head loss h_L due to friction, the required energy input is $175 - 40 = 135.\ ft\ lb_f/lb_m$

$$HP = \frac{100\ lb_m}{min} \times \frac{135\ ft\ lb_f}{lb_m} \times \frac{1}{33{,}000\ ft\ lb_f/(HP\ min)} = 0.41\ HP$$ ●

Answer is (a)

FLUIDS 30

Water flowing in a pipe enters a horizontal venturi tube whose throat area at B is 1/4 that of the original and final cross-sections at A and C.

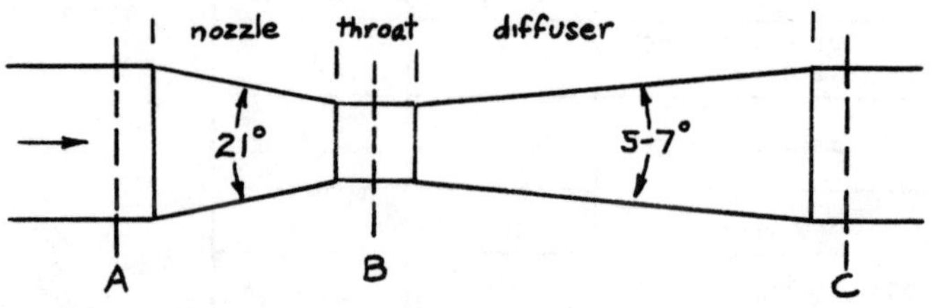

Continuity and energy conservation demand that which one of the following be true?

(a) the pressure at B is increased.
(b) the velocity at B is decreased.
(c) the potential energy at C is decreased.
(d) the flow energy at B is decreased.
(e) kinetic energy at B is reduced.

In a venturi throat the increased velocity, required by continuity, results in a KE increase that occurs at the expense of pressure (flow) energy. ●

Since the system is horizontal, no change in potential energy has occurred. At B the pressure (flow energy) decreases and velocity (KE) increases.

For a well designed venturi, conditions existent at A are essentially restored.

Answer is (d)

FLUIDS 31

Given the following energy data in ft lb_f/lb_m existent at two sections across a pipe transporting water in steady flow:

	Section A	Section B
Potential energy	65	135
Kinetic energy	50	50
Flow energy	335	245
TOTAL	450	430

What frictional head loss in feet has occurred?

(a) 0.
(b) 20.
(c) 70.
(d) 90.
(e) 160.

By energy balance on the basis of 1 lb_m of fluid flowing:

Total energy in = Total energy out + Energy losses - Energy inputs

$$450 = 430 + h_L - 0$$

Head loss, h_L = +20 ft lb_f/lb_m, or 20 feet. ●

Answer is (b)

FLUIDS 32

Theoretical horsepower required to pump water at 100 gallons per minute from a large reservoir to the surface of another large reservoir 400 feet higher is nearest to:

(a) 6. HP
(b) 10. HP
(c) 18. HP
(d) 35. HP
(e) 46. HP

Ignoring frictional losses, pump inefficiency, and any changes in KE or pressure, pumping power is equal to the increase in potential energy given the water.

Z or h = 400 ft lb_f/lb_m potential energy increase per lb_m.

$$\dot{m} = \text{mass flow rate} = 100\ \frac{\text{gal}}{\text{min}} \times 8.33\ \frac{lb_m}{\text{gal}} = 833\ lb_m/\text{min}$$

Basis of calculation: 1 minute

$$HP = \frac{400 \times 833 \text{ ft lb/min}}{33{,}000 \text{ ft } lb_f/(\text{HP min})} = \frac{333{,}200}{33{,}000} = 10.1 \text{ HP}$$ ●

Answer is (b)

FLUIDS 33

A horizontal stream of water with a cross sectional area of 0.1 ft^2 and a velocity of 40. ft/sec has kinetic energy in ft lb_f/sec nearest to:

(a) 2,500.
(b) 6,200.
(c) 11,700.
(d) 56,000.
(e) 132,000.

Volume flow rate, $q = Av = 0.1(40) = 4.0$ cfs

Mass flow rate, $\dot{m} = q\rho = 4.0(62.4 \frac{lb_m}{ft^3}) = 249.6\ lb_m/sec.$

Kinetic energy $= \frac{v^2}{2g}$ ft lb_f/lb_m

Per second: $KE = \left(249.6 \frac{lb_m}{sec}\right)\left(\frac{1600}{64.4}\right) = 6200.$ ft lb_f/sec. ●

Answer is (b)

FLUIDS 34

The theoretical velocity generated by a 10 foot hydraulic head is:

(a) 12.2 ft/sec
(b) 17.9 ft/sec
(c) 25.4 ft/sec
(d) 29.2 ft/sec
(e) 35.8 ft/sec

$V = (2gh)^{1/2} = (2 \times 32.2 \times 10)^{1/2} = (644)^{1/2} = 25.4$ ft/sec. ●

Answer is (c)

FLUIDS 35

What is the static head corresponding to a flow velocity of 10 ft/sec?

(a) 1.55 ft
(b) 1.75 ft
(c) 2.05 ft
(d) 2.25 ft
(e) 2.50 ft

$h = \frac{v^2}{2g} = \frac{10^2}{2(32.2)} = 1.55$ ft. ●

Answer is (a)

FLUIDS 36

Total energy of a compressible or incompressible fluid flowing across any section in a pipeline is a function of:

(a) pressure and velocity.
(b) pressure, density and velocity.
(c) pressure, density, viscosity and velocity.
(d) pressure, density, velocity, and height above datum.
(e) flow energy, kinetic energy, height above datum, and internal energy.

Total energy for a 1st law of thermodynamics energy balance contains all relevant energy terms:

Potential energy, based on height above datum.
Kinetic energy (dynamic or velocity energy), based on velocity.
Flow energy, or pressure energy, based on pressure and density or specific volume.
Internal energy, based on temperature and heat capacity.

The Bernoulli equation for incompressible fluids ignores thermal energy (internal energy), although widely used.

Answer is (e) ●

FLUIDS 37

The locus of the elevations to which water will rise in a piezometer tube is termed:

(a) stagnation pressure
(b) the energy gradient
(c) the hydraulic gradient
(d) friction head
(e) critical depth

Piezometer tube is a static pressure indicating tube, equivalent to a static pressure gage.

Stagnation pressure is an increased pressure developed on impact with a pitot tube as a result of localized kinetic energy reduction to zero.

Hydraulic gradient or hydraulic grade line is flow energy or pressure head in ft lb_f/lb_m plotted vertically above the pipe centerline along the pipe length.

Energy gradient or energy grade line is total energy (flow energy or pressure head, plus kinetic energy or dynamic head, plus potential energy or height above datum) in ft lb_f/lb_m plotted vertically above the elevation datum along the pipe length.

Friction head is head loss h_L in ft lb_f/lb_m under flow conditions.

Critical depth above channel floor in open channels is depth for minimum potential + kinetic energy for the volume flow rate the channel delivers. Tranquil flow (at low KE and high PE) exists above critical depth, and rapid flow (at high KE and low PE) exists below critical depth.

Answer is (c) ●

FLUIDS 38

A water tank consists of a right circular cone that has its central axis vertical and its vertex at the bottom.

The radius at the top is 10 feet, the height is 50 feet, and the water is 15 feet deep.

Determine the work required to pump all the water to the level of the top of the cone.

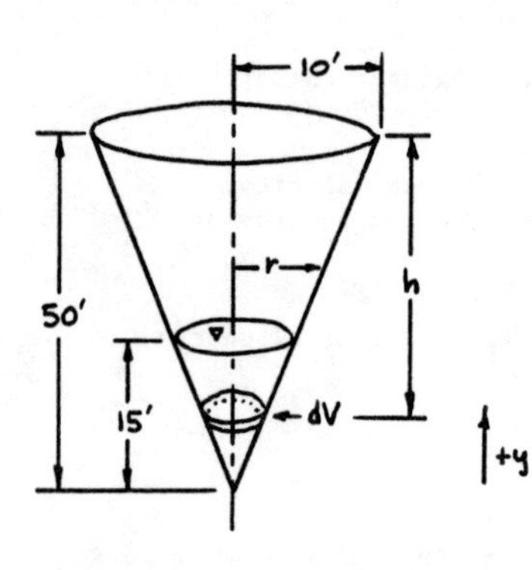

$$W = \rho \int h dV$$

For the cone the differential element of volume is

$$dV = \pi r^2 dy$$

For water $\rho = 62.4$ lbs/ft^3

The radius r is changing by the relation

$$r = \frac{10}{50} y = 0.2y$$

Therefore

$$dV = \pi (0.2y)^2 dy$$

The distance through which the element of volume must be moved $h = (50 - y)$

$$W = 62.4 \int_0^{15} \pi (0.2y)^2 (50 - y) dy$$

$$= 62.4 \pi (0.04) \int_0^{15} 50y^2 - y^3 \, dy$$

$$= 7.83 \left[\frac{50y^3}{3} - \frac{y^4}{4} \right]_0^{15}$$

$$= 7.83 \left[16.67(y^3) - 0.25(y^4) \right]_0^{15}$$

$$= 342 \times 10^3 \text{ ft-lbs.}$$ ●

FLUIDS 39

A stream of fluid with a mass flow rate of 2 slugs/sec and a velocity of 20 ft/sec to the right has its direction reversed 180° in a U fitting. The net dynamic force in lb_f exerted by the fluid on the fitting is nearest to:

(a) 40.
(b) 80.
(c) 514.
(d) 1288.
(e) 2576.

Apply impulse-momentum $Fdt = d(mv)$ in integrated form on the basis of 1 second:

$$F \text{ lbs } (1 \text{ sec}) = (\text{mass flow rate, slugs/sec})(v_2 - v_1, \text{ ft/sec})$$

$$F = 2(-20 - 20) = -80 \text{ lb}_f$$

Since original velocity was taken as +20 in a positive direction, final reversed velocity is -20. Force from impulse-momentum is the force on the fluid necessary to accomplish the velocity change. By reaction, the fluid exerts an equal and opposite force + 80 lb_f to the right on the fitting. ●

Answer is (b)

FLUIDS 40

The thrust in lb_f generated by an aircraft jet engine on takeoff per 1 lb_m/sec of exhaust products, whose velocity has been increased from essentially 0 to 500 ft/sec, is nearest to which of the following:

(a) 15.
(b) 130.
(c) 360.
(d) 710.
(e) 2200.

The combustion products are predominantly nitrogen from the intake air (approx. 15 lb air/lb fuel). Use impulse-momentum:

$$F\,dt = d(mv)$$

Change in momentum = momentum of outlet stream less momentum of air and fuel inlet streams.

$$\Delta(mv) = m_{products}(500) - m_{air}(0) - m_{fuel}(0)$$

Evaluate at 1 second:

$$F = \dot{m}(v_2 - 0) = \left(\frac{1}{32.2}\,\frac{slug}{sec}\right)\left(500\,\frac{ft}{sec}\right) = 15.5\ lb_f \bullet$$

Answer is (a)

FLUIDS 41

A jet of water issues vertically upward from a nozzle with a velocity V ft/sec and a flow rate Q ft^3/sec. Vertically above the nozzle at a distance h feet a horizontal plate is placed as shown.

If the density of the water is w lb/ft^3, what reaction force F is required to keep the plate stationary? Neglect the weight of the plate.

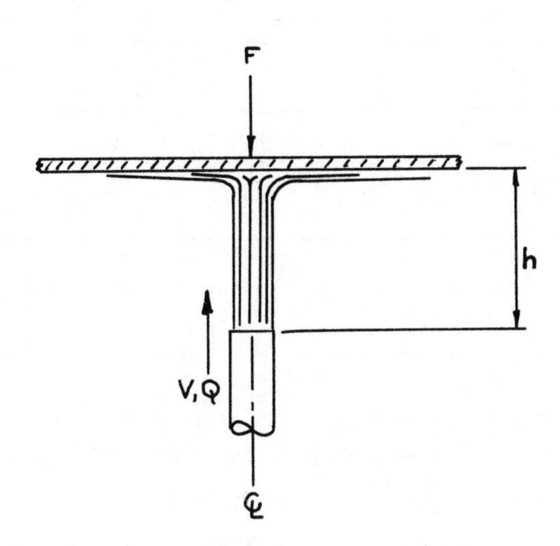

The water squirting upward loses velocity. Use an energy balance to get its velocity just prior to impinging on the plate.

$$\frac{V^2}{2g} - h = \frac{V_2^{\,2}}{2g} \qquad \text{or} \quad V_2 = \sqrt{V^2 - 2gh}$$

Then compute the force F by impulse-momentum for an elapsed time of 1 sec when Q ft^3 flows:

$$F \cdot \cancel{dt}^{\,1} = d(mV) = m dV + V \cancel{dm}^{\,0}$$

$$F\ lb_f(1\ sec) = (Q\ ft^3)\left(\frac{w}{g}\ \frac{lb_f}{ft^3} \cdot \frac{sec^2}{ft}\right)\left(V_2 - 0\ \frac{ft}{sec}\right)$$

$$F = Q\ \frac{w}{g}\ \sqrt{V^2 - 2gh} \quad lbs$$ ●

FLUIDS 42

Which of the following statements most nearly approximates conditions in turbulent flow?

(a) Fluid particles move along smooth, straight paths.
(b) Energy loss varies linearly with velocity.
(c) Energy loss varies as the square of the velocity.
(d) Newton's law of viscosity governs the flow.
(e) The Reynolds number is always less than 2000.

Laminar (streamline, viscous) flow is compared with turbulent flow in the table below.

	Laminar Flow	Turbulent Flow
Motion of fluid particles	Parallel to stream velocity. Paths of particles do not cross.	Particle paths cross and move in all directions.
Energy loss, h_f	$h_f = f\left(\frac{L}{D}\right)\left(\frac{V^2}{2g}\right)$ f is independent of surface roughness and decreases with Re. $f = \frac{64}{Re}$	$h_f = f\left(\frac{L}{D}\right)\left(\frac{V^2}{2g}\right)$ f varies with surface roughness, decreases with Re to constant value. See Moody diagram.
Velocity distribution in pipe	Average is 1/2 of max at centerline. Parabolic distribution. Zero at wall.	Essentially same throughout, except for thin boundary layer at wall. Follows 1/7 power law.
Reynolds number $Re = \frac{DV\rho}{\mu}$	Less than 2000	Greater than 2000

Newtons law of viscosity defines μ on the basis of shear stress and rate of angular deformation. Reynolds number involves μ as one contributing parameter. Very viscous liquids are usually in laminar flow.

On basis of above data, select (c). Do not confuse energy loss, h_f, with friction factor, f.

Answer is (c) ●

FLUIDS 43

For turbulent flow of a fluid in a pipe, all of the following are true, except:

(a) The average velocity will be nearly the same as at the pipe center.
(b) Energy lost due to turbulence and friction varies with kinetic energy.
(c) Pipe roughness affects the friction factor.
(d) Reynolds number will be less than 2000.
(e) Flow of gases and low viscosity liquids is usually turbulent.

In turbulent flow the Reynolds number is greater than 2000.

Answer is (d)

FLUIDS 44

In fluid flow, if the fluid travels parallel to the adjacent layers and the paths of individual particles do not cross, the flow is said to be:

(a) laminar
(b) turbulent
(c) critical
(d) dynamic
(e) uniform

Answer is (a)

FLUIDS 45

Flow nets are drawn to show all of the following except:

(a) Streamlines spaced to bound streamtubes of equal volume flow rate, q.
(b) Paths of particles in 2-and 3-dimensional flow.
(c) Equipotential lines perpendicular to streamlines.
(d) Equipotential lines spaced equal to distance between the streamlines they cross.
(e) Flows and pressure drops in distribution systems.

All are true except (e). Pipe networks (containing parallel pipes, different size pipes in series, and branch pipes) are analyzed analogously to electrical networks.

Typical streamlines and equipotential lines, existent in 2-dimensional flow in the bend of a passage, are shown in the drawing.

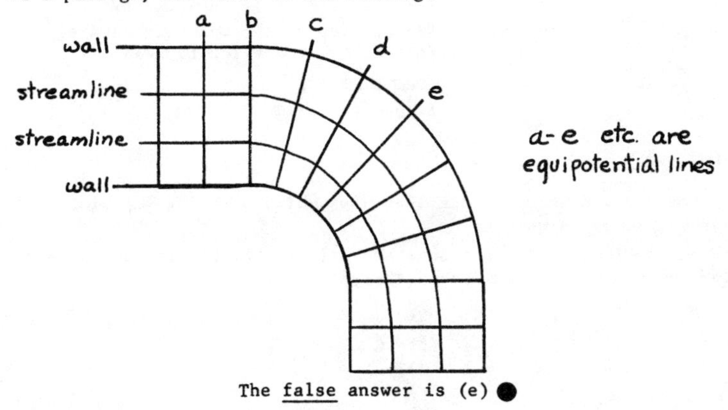

The false answer is (e)

FLUIDS 46

Which of the following constitutes a group of parameters with the dimensions of power?

(a) ρAV

(b) PAV

(c) $\frac{DV\rho}{\mu}$

(d) $\frac{\rho V^2}{P}$

(e) $\frac{V^2}{Dg}$

(a) is mass flow rate, m in slugs/sec or lb_m/sec.

(b) Power has dimensions of ft lb_f/sec. PAV has these dimensions.

(c) is Reynolds number, Re, dimensionless, ratio of inertial force to viscous force.

(d) is Euler number, Eu, dimensionless, ratio of inertial force to pressure force.

(e) is Froude number, Fr, dimensionless, ratio of inertial force to gravity force.

Parameters involved are:

A, cross sectional area, ft^2

D, diameter, ft

P, pressure, lb_f/ft^2

V, velocity, ft/sec

μ = absolute viscosity, $lb_f\ sec/ft^2$ or slug/(ft sec)

ρ = density, $lb_f\ sec^2/ft^4$ or lb_m/ft^3 or $slug/ft^3$

slug = $lb_f\ sec^2/ft = \frac{lb_m}{g_c}$

g = acceleration due to gravity 32.17 ft/sec^2

g_c = conversion factor = 32.17 $lb_m\ ft/(lb_f\ sec^2)$

Answer is (b) ●

FLUIDS 47

At or below critical velocity in small pipes or at very low velocities, the loss of head due to friction

(a) varies directly as velocity.
(b) can be ignored.
(c) is infinitely large.
(d) varies as the velocity squared.
(e) equals the velocity head.

Answer is (a) ●

FLUIDS 48

Given the Moody diagram on the next page which is a log-log plot of friction factor vs. Reynolds number.

Which of the lines A-E represents the friction factor to use for a smooth pipe of low roughness ratio (ϵ/D) in streamline flow?

(a) A
(b) B
(c) C
(d) D
(e) E

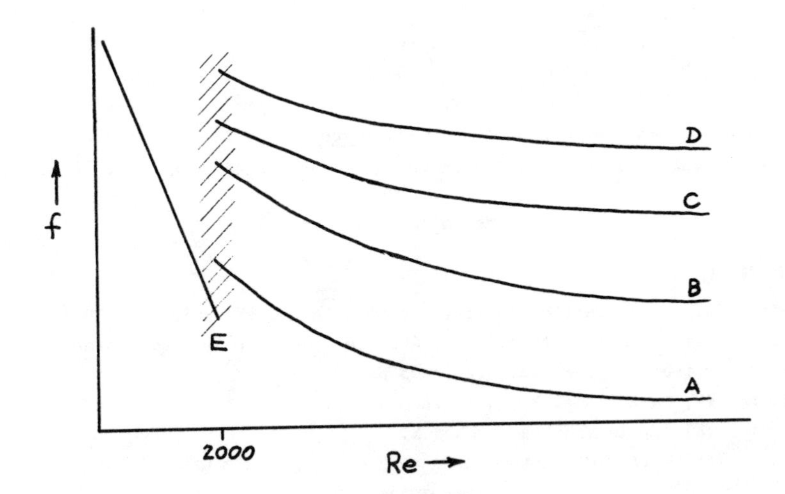

Line E applies to all roughness ratios in streamline flow (Re < 2000) because stagnant fluid layers at the wall make friction factor independent of roughness ratio; $f = \frac{64}{Re}$. In turbulent flow (Re > 2000) increasing roughness is represented by A for smooth pipe, to D for very rough pipe. In turbulent flow, where only the thinnest boundary layer exists, friction factor is very dependent on surface roughness.

Answer is (e) ●

FLUIDS 49

At constant pressure P all of the following statements about flow rate, q in turbulent flow in a pipe are true, except:

(a) q varies approximately as (diameter, D)^2.

(b) q decreases as pipe surface roughness ratio, $\frac{\epsilon}{D}$ increases.

(c) q decreases as pipe length, L increases.

(d) q decreases inversely as (fluid density, ρ)$^{1/2}$

(e) q varies as (velocity, V)^2

At steady state, an energy balance equates flow (pressure) energy with frictional loss + KE. $\frac{P}{\rho} = h_L + \frac{V^2}{2g_c}$.

The Darcy-Weisbach equation quantifies loss: $h_L = f\left(\frac{L}{D}\right)\frac{V^2}{2g_c}$. The friction factor f is a function of Reynolds no., Re, and surface roughness, $\left(\frac{\epsilon}{D}\right)$ as shown on the Moody diagram of the previous question. In turbulent flow, f usually ranges from 0.01 to 0.06 and becomes constant at high Re. Since q = AV, an overall expression for flow can be assembled for circular pipes in terms of the parameters D, P, ρ, L and f. Then f can be determined from $Re = \frac{DV\rho}{\mu}$ and $\frac{\epsilon}{D}$.

$$q = \frac{\pi}{4} D^2 \left[\frac{2g_c P}{\rho \left[1 + f \left(\frac{L}{D}\right)\right]} \right]^{1/2}$$

All of the problem statements are correct, except (e), for q = AV shows that q varies directly as velocity. Note that the derived equation shows that q varies as $P^{1/2}$

The false answer is (e) ●

FLUIDS 50

A 24-inch water pipe carries 15 cfs. At point A the elevation is 150 feet and the pressure is 30 psig. At point B, 4000 feet downstream from A, the elevation is 130 feet and the pressure is 35 psig.

Determine (a) head loss, h_L, between A and B.
(b) HP loss between A and B.
(c) friction factor, f.

(1) Use energy balance to determine h_L. Basis: 1 lb_m of fluid.

PE + KE + Pressure (flow) energy = PE + KE + Pressure (flow) energy + h_L

$$150 + \frac{V^2}{2g_c} + \frac{30(144)}{62.4} = 130 + \frac{V^2}{2g_c} + \frac{35(144)}{62.4} + h_L$$

Since the pipe diameter is unchanged, continuity requires that V be the same at both points. KE can be deleted from both sides of the equation.

$$150 + 69.2 = 130 + 80.8 + h_L$$

$$h_L = 8.40 \quad \text{ft } lb_f/lb_m$$

(2) Calculate energy loss in HP. Basis: 1 second

$$\text{mass flow rate, } \dot{m} = q\rho = \left(15 \frac{ft^3}{sec}\right)\left(62.4 \frac{lb_m}{ft^3}\right) = 936.\ lb_m/sec$$

$$\text{energy loss} = \left[936 \frac{lb_m}{sec}\right]\left[8.40 \frac{ft\ lb_f}{lb_m}\right] = 7862. \frac{ft\ lb_f}{sec}$$

$$\text{HP loss} = \frac{7862 \quad ft\ lb/sec}{550\ ft\ lb/(HP\ sec)} = 14.3\ HP$$

(3) Calculate velocity, V, and substitute in Darcy-Weisbach equation.

$q = AV$ $\quad A = \frac{\pi}{4}(2)^2 = \pi\ ft^2$ $\quad V = \frac{15}{\pi} = 4.78$ ft/sec

$$h_L = f\left(\frac{L}{D}\right)\frac{V^2}{2g_c} \qquad 8.40 = f\left[\frac{4000}{2}\right]\frac{22.85}{2(32.17)}$$

$$f = 0.0118$$ ●

FLUIDS 51

Entrance losses between tank and pipe, or losses through elbows, fittings and valves are generally expressed as functions of

(a) kinetic energy
(b) pipe diameter
(c) friction factor
(d) volume flow rate
(e) all of the above

Typical head losses for the above items are expressed as an empirical average constant, K, times kinetic energy, $\frac{v^2}{2g_c}$. $h_L = K \frac{v^2}{2g_c} = \text{ft } lb_f/lb_m$.

Values of K vary widely, but the list below is typical for minor losses:

	K
Entrance loss - flush with wall	0.50
- projecting inside tank	1.0
- smoothy contoured	0.05
Exit loss - entering tank	1.0
90° Elbow	0.90
Gate valve - fully open	0.20
Globe valve - fully open	10.0

Answer is (a) ●

FLUIDS 52

For the diagram shown, compute the velocity of the water flowing in the 1000' branch of the 6" diameter pipe.

Assume the friction factors in the two 6" round pipes are the same and that the incidental losses are equal in the two branches.

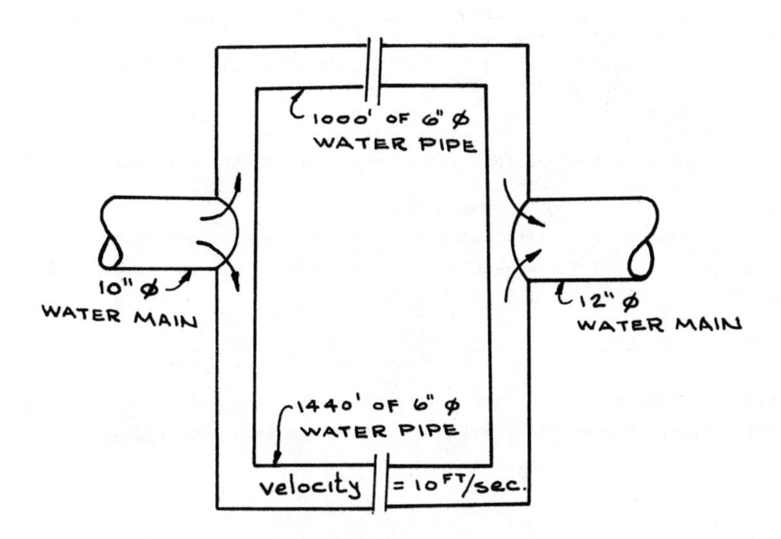

There is a drop in the energy line from the 10" main to the 12" main. This head loss must be equal in both 6" branches.

$$h_{L1000'} + \text{incidental losses}_{1000'} = h_{L1440'} + \text{incidental losses}_{1440'}$$

The Darcy equation is

$$h_L = f \frac{L}{d} \frac{V^2}{2g}$$

where h_L = head loss in feet
f = friction factor
L = length of pipeline in ft.
d = diameter of pipe in ft.
g = 32.2 ft/sec^2

Thus in this situation:

$$f \frac{1000'}{0.5'} \frac{V_{1000}^2}{2g} = f \frac{1440'}{0.5'} \frac{V_{1440}^2}{2g}$$

which reduces to

$$1000V^2_{1000} = 10^2(1440)$$

$$V_{1000'} = \sqrt{\frac{100(1440)}{1000}} = 12 \text{ ft/sec}$$ ●

FLUIDS 53

The vena contracta of a sharp edged hydraulic orifice usually occurs:

(a) At the geometric center of the orifice.
(b) At a distance of about 10% of the orifice diameter upstream from the plane of the orifice.
(c) At a distance equal to about two orifice diameters downstream from the plane of the orifice.
(d) At a distance equal to about one-half the orifice diameter downstream from the plane of the orifice.
(e) At a distance within 10% of the orifice diameter downstream from the plane of the orifice.

The vena contracta is located about 1/2 orifice diameter downstream from the orifice. The reduced cross sectional area of the stream is expressed as a coefficient of contraction, $C_c = \frac{\text{area at vena contracta}}{\text{area of orifice}}$.

A coefficent of velocity, $C_v = \frac{\text{actual velocity at v.c.}}{\text{theoretical velocity at v.c.}}$.

Theoretical velocity at the vena contracta ignores the existence of an entrance loss to the orifice, and for a hole in a tank equals $\sqrt{2g_c h}$.

Coefficient of discharge, $C_D = C_c C_v$.

Answer is (d) ●

FLUIDS 54

An orifice 2" in diameter discharges fluid from a tank with a head of 15 feet. Discharge rate, q, is measured at 0.5 cfs. Actual velocity at the vena contracta, v.c., is 29.0 ft/sec. The coefficient of discharge, C_D, is nearest to:

(a) 0.62
(b) 0.74
(c) 0.79
(d) 0.86
(e) 0.94

$C_D = C_c C_v$ where C_D = coefficient of discharge

C_c = coeff. of contraction = $\dfrac{\text{area of v.c.}}{\text{area of orifice}}$

C_v = coeff. of velocity = $\dfrac{\text{actual velocity at v.c.}}{\text{theoretical velocity at v.c.}}$
(ignoring losses)

Theoretical velocity at v.c. = $V = \sqrt{2g_c h} = \sqrt{2(32.17)(15)} = 31.1$ ft/sec

$$C_v = \frac{29.0}{31.1} = 0.935$$

Area of v.c. $\quad Q = AV \quad A = \dfrac{Q}{V} = \dfrac{0.5}{29} = 0.0173 \text{ ft}^2$

Area of orifice $= \dfrac{\pi}{4} D^2 = \dfrac{\pi}{4}\left(\dfrac{1}{6}\right)^2 = \dfrac{\pi}{144} = 0.0218 \text{ ft}^2$

$$C_c = \frac{0.0173}{0.0218} = 0.794$$

$$C_D = C_c C_v = (0.794)(0.935) = 0.742$$

Answer is (b)

FLUIDS 55

A stream of incompressible fluid leaves a nozzle at an absolute velocity of 50 ft/sec to the right. It enters, traverses and leaves tangent to a quarter circular blade which is moving at 30 ft/sec to the right as shown in the drawing. The magnitude of its absolute velocity in ft/sec after leaving the blade is nearest to:

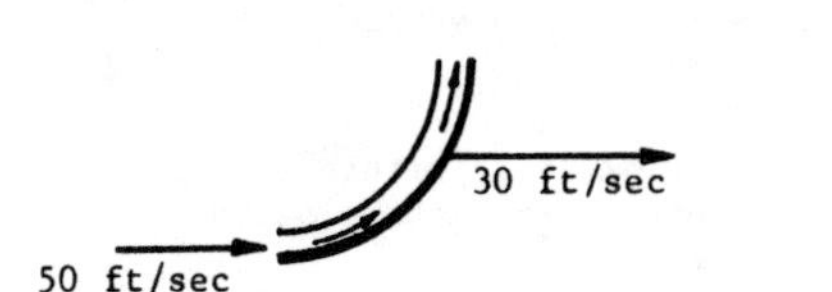

(a) 24.
(b) 30.
(c) 36.
(d) 42.
(e) 48.

Ignoring frictional losses and any change of cross sectional area, fluid enters in line with blade travel at 20 ft/sec relative to the blade, and leaves normal to blade travel at 20 ft/sec relative to the blade.

Since the fluid also moves in the direction of blade motion at blade velocity, its absolute velocity as shown in the vector triangle is 36 ft/sec. ●

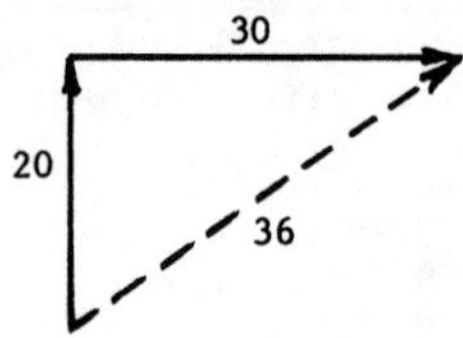

Fluid relative velocity + blade absolute velocity = fluid absolute velocity

Answer is (c)

FLUIDS 56

At normal atmospheric pressure, the maximum height in feet that a non volatile fluid of specific gravity of 0.80 may be siphoned is nearest to:

(a) 12.
(b) 20.
(c) 28.
(d) 34.
(e) 42.

Maximum height to which a fluid may be siphoned is determined when pressure of the fluid column plus its vapor pressure equals external pressure. Minimum pressure at the highest point is 0 psia plus vapor pressure.

A 34 foot column of water has a pressure of 14.7 psi; this ignores its vapor pressure that is equivalent to 0.8 feet at 68 F.

$$(\text{ft of fluid})(62.4 \text{ lb/ft}^3 \times 0.80) = 14.7 \times 144 \text{ lb/ft}^2$$

$$\text{ft of fluid} = \frac{14.7 \times 144}{62.4 \times 80} = 42.4$$ ●

Answer is nearest to (e)

FLUIDS 57

Cavitation is the result of

(a) static pressure in a fluid becoming less than fluid vapor pressure
(b) rivets under impact load
(c) exposure of concrete to salt water
(d) heat treatment of a low carbon steel
(e) improper welding technique

Cavitation is characterized by local reduction of pressure within a fluid to the vapor pressure and the formation of a cavity within the flowing fluid. ●

Answer is (a)

FLUIDS 58

On doubling the speed, N, of a centrifugal pump, all of the following are true, except:

(a) Head, H is increased by a factor of 4.
(b) Horsepower, P is increased by a factor of 8.
(c) Volume flow rate, Q is increased by a factor of 2.
(d) Torque input is increased by a factor of 4.
(e) Head, horsepower and volume flow rate are independently variable.

Similarity relations are used between model and a geometrically similar prototype whose impeller diameters, D differ. These relations are:

Speed: $$\frac{DN}{H^{1/2}} = \text{constant} \qquad \frac{NQ^{1/2}}{H^{3/4}} = \text{constant, specific speed}$$

Discharge: $$\frac{Q}{D^3N} = \text{constant} \qquad \frac{Q}{D^2H^{1/2}} = \text{constant}$$

Power: $$\frac{P}{D^5N^3} = \text{constant}$$

Applied to a single pump of fixed impeller diameter D, these reduce to

$$\frac{N_1}{N_2} = \frac{Q_1}{Q_2} = \frac{(H_1)^{1/2}}{(H_2)^{1/2}} = \frac{(P_1)^{1/3}}{(P_2)^{1/3}}$$

Doubling speed concurrently increases Q by a factor of 2,
increases H by a factor of 4, and
increases P by a factor of 8.

Mutual dependence of these variables complicates selection of pumps for specific applications.

The false answer is (e) ●

FLUIDS 59

Water flow rate in a 6 inch diameter pipe is measured with a differential pressure gage connected between a static pressure tap in the pipe wall and a pitot tube located at the pipe centerline. Which volume flow rate, q in cfs, results in a differential pressure of 1 psi?

(a) 0.2
(b) 2.4
(c) 15.6
(d) 28.2
(e) 344.0

A pitot tube generates a stagnation pressure as fluid kinetic energy is converted to pressure head.

Basis of calculation: 1 lb_m of fluid flowing

$$\frac{V^2}{2g_c} = \frac{P}{\rho} \qquad V = \left[\frac{2g_cP}{\rho}\right]^{1/2}$$

$$V = \left[\frac{2(32.17 \frac{lb_m\ ft}{lb_f\ sec^2})(1 \frac{lb_f}{in^2})(\frac{144\ in^2}{ft^2})}{62.4 \frac{lb_m}{ft^3}} \right]^{1/2} = 12.2\ ft/sec$$

$$q = AV = \frac{\pi}{4} D^2 V = \frac{\pi}{4} \left(\frac{1}{2}\right)^2 (12.2) = 2.39\ ft^3/sec \bullet$$

Answer is (b)

FLUIDS 60

The hydraulic formula $CA\sqrt{2gh}$ is used to find the

(a) quantity of discharge through an orifice
(b) velocity of flow in a closed conduit
(c) length of pipe in a closed network
(d) friction factor of a pipe
(e) wetted perimeter of a canal

For a static head orifice discharging freely into the atmosphere

$$Q = CA\sqrt{2gh} \bullet$$

Answer is (a)

FLUIDS 61

The hydraulic radius of an open-channel section is defined as:

(a) the wetted perimeter divided by the cross sectional area.
(b) the cross sectional area divided by the total perimeter.
(c) the cross sectional area divided by the wetted perimeter.
(d) one-fourth the radius of a circle with the same area.
(e) four times the radius of a circle with the same area.

Hydraulic radius R, in feet, is defined as $\frac{\text{cross sectional area, ft}^2}{\text{wetted perimeter, ft}} \bullet$

For a full circular pipe: $R = \frac{\pi r^2}{2\pi r} = \frac{r}{2} = \frac{D}{4}$

Answer is (c)

FLUIDS 62

To calculate a Reynolds number for flow in open channels and in cross sections, one must utilize hydraulic radius, R in feet and modify the usual expression for circular cross sections:

$$Re = \frac{DV\rho}{\mu} = \frac{VD}{\nu}$$

where
D = diameter in feet
V = velocity in ft/sec
ρ = density in slugs/ft^3
μ = absolute viscosity in slugs/(ft sec)
ν = kinematic viscosity in ft^2/sec

Which of the following modified expressions for Re is applicable to flow in open or non-circular cross sections?

(a) $\frac{RD}{\nu}$

(b) $\frac{RV\rho}{\mu}$

(c) $\frac{2RD}{\nu}$

(d) $\frac{2RV\rho}{\mu}$

(e) $\frac{4RV}{\nu}$

Choices (a) and (c) are not dimensionless as required for Reynolds number.

Since hydraulic radius $R = \frac{\text{cross sectional area}}{\text{wetted perimeter}}$,

For a circular cross section:

$$R = \frac{\frac{\pi D^2}{4}}{\pi D} = \frac{D}{4}$$

Therefore,

$$Re = \frac{4RV\rho}{\mu} = \frac{4RV}{\nu}$$ ●

Answer is (e)

FLUIDS 63

In the design of waterways, the "hydraulic jump" is sometimes used for

(a) energy dissipation
(b) elimination of turbulence
(c) prevention of sedimentation
(d) measurement of flow
(e) reduction of head loss

Answer is (a) ●

FLUIDS 64

The rate of laminar water flow in a saturated soil can be calculated using:

(a) a Moody diagram.
(b) the Bernoulli equation.
(c) Darcy's law.
(d) the Hazen-Williams formula.
(e) the Chezy or Manning equations.

A Moody diagram graphically presents the relationship for pipes among the dimensionless parameters: friction factor (f), Reynolds no. (Re), and surface roughness, $\frac{\epsilon}{D}$

The Bernoulli equation is a mechanical energy balance based on the 1st law of thermodynamics. It includes PE, KE, flow energy, losses and shaft work inputs, but omits thermal energy. The Euler equation is similar, though written in differential equation form.

The Darcy-Weisbach formula relates head loss (h_L) with friction factor (f), length-diameter ratio $\frac{L}{D}$ and kinetic energy $\frac{v^2}{2g}$.

$$h_L = f \left(\frac{L}{D}\right) \frac{v^2}{2g}$$

Darcy's law for diffusional flow closely parallels conductive heat transfer. As applied in this case:

$$Q = kiA$$

where Q = volume flow rate
k = coefficient of permeability
i = gradient
A = cross sectional area

The Hazen-Williams formula is an empirical equation used in the solution of piping networks:

$$V = 1.318 C_1 R^{0.63} S^{0.54}$$

where V = velocity in ft/sec
C_1 = a coefficient of relative roughness
R = hydraulic radius in ft
S = slope of the hydraulic gradient in ft lb_f/lb_m per ft length.

The Chezy formula applies to steady flow in open channels:

$$V = C\sqrt{RS}$$

where V = velocity
C = an empirical coefficient
R = hydraulic radius
S = slope of the hydraulic gradient (or slope of the free water surface in steady flow).

The Manning equation similarly applies to steady flow in open channels:

$$Q = AV = \frac{1.486}{n} AR^{2/3} S^{1/2}$$

where Q = volume flow rate
A = cross sectional area
n = surface roughness factor
R = hydraulic radius
S = slope of the hydraulic gradient, as above.

Answer is (c) ●

FLUIDS 65

A rain storm occurred over a 100 square mile drainage area located in the Central Valley. The storm had a total duration of 3.0 hours and amounted to a total average rainfall over the drainage area of 1.00 inches of which 0.50 inch fell in the first hour, 0.30 inch in the second hour, and 0.20 inch in the third or last hour. Soil type A occurs in 30% of the area, Soil type B occurs in 25% of the area, and Soil type C occurs in 45% of the area.

The following data give the rainfall infiltration rates applicable to this 100 square mile drainage area:

	Rainfall Infiltration Rate*		
	Soil type A	Soil type B	Soil type C
First Hour	.40"	.35"	.30"
Second Hour	.25"	.20"	.15"
Third Hour	.20"	.15"	.10"

*Inches of rainfall absorbed by the soil per hour

Using the above data, compute the volume of storm runoff (cubic feet) that would originate from this 100 square mile drainage area.

A tabular form simplifies the computation:

	Total Rain fall	Soil Type A 30 Sq Miles Infl.	Soil Type A 30 Sq Miles Run-off	Soil Type B 25 Sq Miles Infl.	Soil Type B 25 Sq Miles Run-off	Soil Type C 45 Sq Miles Infl.	Soil Type C 45 Sq Miles Run-off
First Hour	0.5	0.40	0.10	0.35	0.15	0.30	0.20
Second Hour	0.3	0.25	0.05	0.20	0.10	0.15	0.15
Third Hour	0.2	0.20	0	0.15	0.05	0.10	0.10
	1.0		0.15		0.30		0.45

Runoff = 30(0.15) + 25(0.30) + 45(0.45) = 32.25 sq mile - inches

Convert square miles to square feet and inches to feet

$$\text{Runoff} = 32.25 \times (5{,}280)^2 \times (1/12) = 75 \times 10^6 \text{ cubic feet}$$ ●

FLUIDS 66

A concrete pipe is laid on a slope of 0.0001 foot per foot and is to carry 52.8 cubic feet per second of water when the pipe flows 0.75 diameter full.

With the use of Manning's formula and an "n" factor of 0.015, what size pipe should be used?

The Manning formula:

$$Q = \frac{1.486}{n} AR^{2/3} S^{1/2}$$

where n = roughness factor (0.015)
S = slope (0.0001)
Q = flow rate (52.8 cu ft/sec)
A = area of flow
R = hydraulic radius

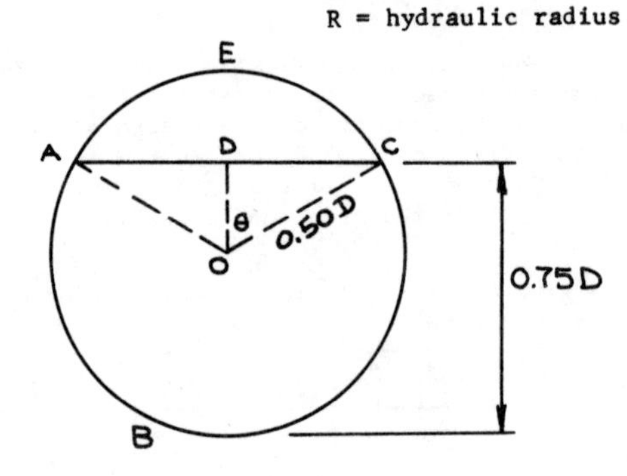

$$\text{Hydraulic radius } R = \frac{A}{P} = \frac{\text{Area of flow}}{\text{Wetted perimeter}}$$

$$= \frac{\text{Area of circle - (Sector AOCE - Triangle AOC)}}{\text{Arc ABC}}$$

$$\text{Angle } \Theta = \cos^{-1}\left(\frac{0.25D}{0.50D}\right) = \cos^{-1} 0.5 \qquad \Theta = 60°$$

$$\text{Area of Sector AOCE} = 2\left(\frac{60°}{360°}\right)\frac{1}{4}\pi D^2 = 0.262D^2$$

$$\text{Length of Arc ABC} = \pi D - 2\left(\frac{60°}{360°}\right)\pi D = 2.09D$$

$$\text{Area of Triangle AOC} = \frac{1}{2}(0.866)\left(\frac{D}{2}\right)(2)\left(\frac{D}{4}\right) = 0.108D^2$$

$$R = \frac{A}{P} = \frac{0.785D^2 - (0.262D^2 - 0.108D^2)}{2.09D} = \frac{0.631D^2}{2.09D} = 0.302D$$

Substitute the values into the Manning equation

$$52.8 = \frac{1.486}{0.015}(0.631D^2)(0.302D)^{2/3}(0.0001)^{1/2}$$

$$D^{8/3} = 187.5$$

$$D = 7.12 \text{ feet}$$

7

Thermodynamics

THERMO 1

Equations of state for a single component can be any of the following, except:

(a) the ideal gas law, Pv = RT.
(b) the ideal gas law modified by insertion of a compressibility factor, Pv = ZRT.
(c) any relationship interrelating 3 or more state functions.
(d) a mathematical expression defining a path between states.
(e) relationships mathematically interrelating thermodynamic properties of the material.

All except (d) are correct. The ideal gas law is the simplest equation of state; it is often applied to real gases by using a compressibility factor Z. Any relationships that interrelate thermodynamic state function data are equations of state. Item (d) expresses the path of a process between states rather than a relationship between variables at a single point or state.

Answer is (d) ●

THERMO 2

The state of a thermodynamic system is always defined by its

(a) absolute temperature
(b) process
(c) properties
(d) temperature and pressure
(e) availability

Availability is the amount of work that exists above complete equilibrium with ambient conditions. State is always defined by its properties. The number of properties required is determined by the Gibbs phase rule $P + F = C + 2$, where P = no. of phases, F = degrees of freedom or number of properties required, and C = no. of distinguishable components present.

Answer is (c) ●

THERMO 3

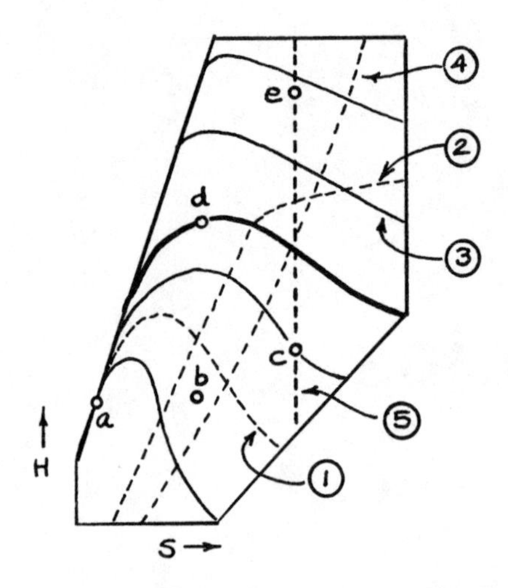

On the Mollier diagram for steam, which of the numbered lines represents a line of constant pressure?

(a) Line 1
(b) Line 2
(c) Line 3
(d) Line 4
(e) Line 5

(a) constant moisture 100 - x% or quality x%. (Line 1)

(b) constant temperature, °F. (Line 2)

(c) constant superheat, °F. (Line 3)

(d) constant pressure, psia. (Line 4)

(e) constant entropy, BTU/lb°R. (Line 5)

Horizontal lines are constant enthalpy or total heat content.

Answer is (d)

THERMO 4

On the Mollier diagram for steam (above) which lettered point represents the critical point at 3206 psia and 705°F?

The following data can be read for these points on a large Mollier diagram:

	Temp °F	P psia	Remarks	H BTU/lb	S BTU/lb°F
a	705	3206	critical point	910	1.065
b	162	5	22% moisture	910	1.490
c	141	3	10% moisture	1010	1.760
d	440	375	dry saturated	1205	1.490
e	780	200	superheated 400°	1413	1.760

Answer is (a)

THERMO 5

Mathematically, a thermodynamic property is which of the following?

(a) a point function
(b) a path function
(c) an inexact differential
(d) discontinuous
(e) an exact differential

Thermodynamic properties, such as P, v, T, U, H, and S, are point functions, their derivatives are exact differentials. Heat, Q, and work, W, are path functions expressible as areas that are dependent upon path between points (states), their derivatives are inexact differentials.

Answer is (a)

THERMO 6

T	P psia	v_f	v_g	h_f	h_g	S_f	S_g
40°F	51.7	0.0116	0.774	17.3	81.4	0.0375	0.166
80°F	98.9	0.0123	0.411	26.4	85.3	0.0548	0.164
120°F	172.	0.0132	0.233	36.0	88.6	0.0717	0.162

Given the above data for Freon 12, what is its state at 40°F and 25 psia?

(a) saturated liquid
(b) superheated vapor
(c) compressed liquid
(d) saturated vapor
(e) vapor-liquid mixture

At 40°F equilibrium between liquid and gas exists at 51.7 psia. Below 51.7 psia superheated vapor exists, and above 51.7 psia only pressurized liquid exists.

Answer is (b) ●

THERMO 7

Using the previous Freon 12 data table, what is its entropy in BTU/lb°R at 120°F and 80% quality?

(a) 0.057
(b) 0.144
(c) 0.186
(d) 28.8
(e) none of these

At 120°F, $S_f = 0.0717$ and $S_g = 0.162$. $S_{fg} = 0.162 - 0.0717 = 0.090$

S_f is saturated liquid at 0% quality and S_g is saturated vapor of 100% quality.

S at 80% quality = $S_f + (0.80)S_{fg} = 0.0717 + (0.80)(0.090) = 0.144$ BTU/lb°R. ●

Answer is (b)

THERMO 8

Using the previous Freon 12 data table, what is its latent heat (heat of vaporization) in BTU/lb at 80°F?

(a) 0.219
(b) 0.423
(c) 26.4
(d) 58.9
(e) none of these

$h_{fg} = h_g - h_f = 85.3 - 26.4 = 58.9$ BTU/lb ●

Answer is (d)

THERMO 9

Normal boiling point of liquid oxygen is 90°K. What is this temperature in °R?

(a) - 330°R
(b) - 183
(c) 162
(d) 168
(e) 518

$(\frac{1.8°R}{°K})(90°K) = 162°R.$ ●

Answer is (c)

THERMO 10

For a typical refrigerant, draw a pressure-enthalpy diagram which shows the boundaries of solid, liquid, and vapor states. Identify on the diagram the following:
(a) superheated vapor zone
(b) critical pressure
(c) compressed liquid zone
(d) saturated liquid line

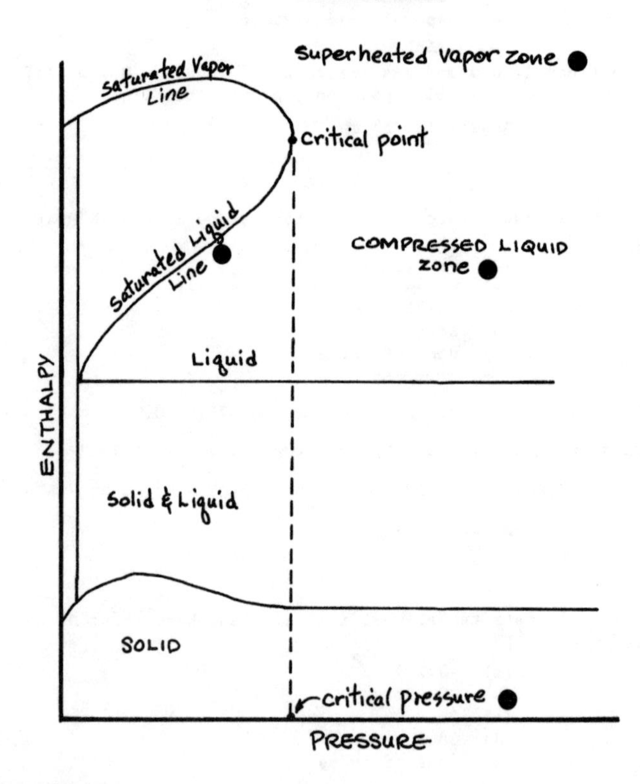

THERMO 11

A cylinder fitted with a weightless, frictionless piston contains m pounds of air at temperature T_1, volume V_1, and ambient pressure P_a. Heat is added until the air in the cylinder has a temperature T_2, a volume V_2, and ambient pressure P_a. The specific heat of air at constant pressure is C_p, and the specific heat of air at constant volume is C_v. The heat transferred during the process is:

(a) $mC_p(T_2 - T_1)$
(b) $mC_v(T_2 - T_1) - P_a(V_2 - V_1)$
(c) $mC_p(T_2 - T_1) + P_a(V_2 - V_1)$
(d) $mC_v(T_2 - T_1) + P_a(V_2 - V_1)$
(e) $mC_v(T_2 - T_1)$

Consider one or more processes from $P_aV_1T_1$ to $P_aV_2T_2$ in a closed system and apply the first law of thermodynamics: $\Delta U = Q - W$.

For an ideal gas (air): $\Delta U = mC_v \Delta T$

Work done by a closed system at constant pressure: $W = \int_{V_1}^{V_2} P dV = P_a(V_2 - V_1)$

Combine these: $mC_v(T_2 - T_1) = Q - P_a(V_2 - V_1)$

Therefore $Q = mC_v(T_2 - T_1) + P_a(V_2 - V_1)$ ● Answer is (d)

Dimensionality check:
Compatible units would be

$$BTU = (\text{lb mole})\left(\frac{BTU}{°F\ \text{lb mole}}\right)(°F) + \left(\frac{\text{lbf}}{\text{ft}^2}\right)(\text{ft}^3)\left(\frac{BTU}{778\ \text{lbf ft}}\right)$$

THERMO 12

A nonflow (closed) system contains 1 lb of an ideal gas ($C_p = 0.24$, $C_v = 0.17$). The gas temperature is increased by 10°F while 5 BTU of work are done by the gas. What is heat transfer in BTU?

(a) -3.3
(b) -2.6
(c) +6.7
(d) +7.4
(e) none of these

The thermodynamic sign convention is + for heat in and + for work out of a system. Apply the first law for a closed system and an ideal gas working fluid:

$$\Delta U = mC_v \Delta T = Q - W$$

$0.17(10) = Q - (+5)$ $1.7 = Q - 5$, $Q = 6.7$ ●

Answer is (c)

THERMO 13

Shaft work of -15 BTU/lb and heat transfer of -10 BTU/lb change enthalpy of a system by

(a) -25 BTU/lb
(b) -15 BTU/lb
(c) -10 BTU/lb
(d) -5 BTU/lb
(e) +5 BTU/lb

The first law applied to a flow system is:

$$H = Q - W_s = -10 - (-15) = +5$$ ●

Answer is (e)

THERMO 14

The first law of thermodynamics states that:

(a) heat energy cannot be completely transformed into work.
(b) internal energy is due to molecular motion.
(c) heat can only be transferred from a body of higher termperature to one of lower temperature.
(d) energy can be neither created nor destroyed.
(e) entropy of the universe is increased by irreversible processes.

Item (a) is a consequence of the second law. Item (b) is a statement derived from the fact that internal energy is thermal kinetic energy of the molecules.

Item (c) is untrue as stated because a heat pump or refrigerator transfers heat from a low temperature reservoir to a higher temperature. Item (d) is correct, excluding nuclear processes that convert mass to energy. The first law is often cited as "conservation of energy." Item (e) is true, and is a consequence of the second law.

Answer is (d) ●

THERMO 15

A thermodynamic system originally contains a rock on a cliff and a pond of water below, both at the same temperature as the surroundings. The rock originally with PE = 4 BTU falls into the water and both ultimately come again into the same thermal equilibrium with the surroundings. Which row of data best represents the NET change in BTU that the rock plus pond system has undergone?

	ΔU	ΔKE	ΔPE	Q	W
(a)	0	0	-4	-4	0
(b)	+4	0	-4	0	0
(c)	0	+4	-4	0	0
(d)	0	0	0	-4	+4
(e)	-4	+4	-4	+4	0

Define the system as rock plus pond, with all else as surroundings. Initially rock has PE = +4 and KE = 0, and pond has PE = 0 and KE = 0, for a total = +4.

Finally rock has PE = 0 and KE = 0, and pond has PE = 0 and KE = 0, for a total of 0.

Energy was dissipated as heat, -Q to the surroundings. No net temperature change has occurred so ΔU = 0. No work was done by the system so W = 0.

Net result: ΔU = 0, ΔKE = 0, ΔPE = -4, Q = -4 and W = 0. ●

Answer is (a)

THERMO 16

A fluid at 100 psia has a specific volume of 4 ft^3/lb. and enters an apparatus with a velocity of 500 ft/sec. Heat radiation losses in the apparatus are equal to 10 BTU/lb. of fluid supplied. The fluid leaves the apparatus at 20 psia with a specific volume of 15 ft^3/lb. and a velocity of 1000 ft/sec. In the apparatus, the shaft work done by the fluid is equal to 195,000 ft-lb_f/lb_m.

Does the internal energy of the fluid increase or decrease, and how much is the change?

Basis of calculation: 1 lb_m

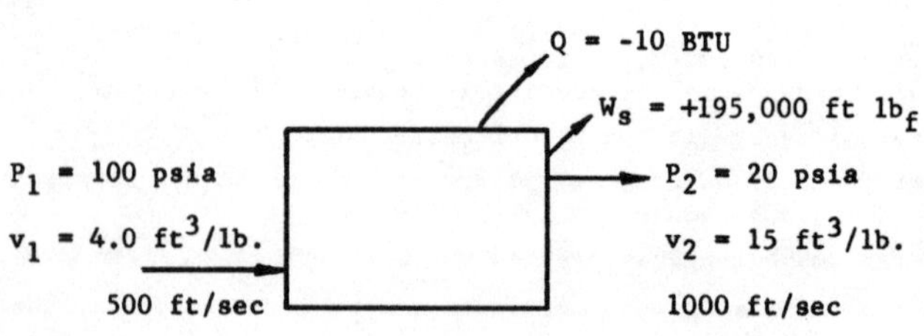

Use the thermodynamic sign convention that heat in and work out are positive.

First law energy balance for the flow system:

$$h_2 + KE_2 - h_1 - KE_1 = Q - W_s$$

Since the working fluid is unspecified and internal energy change is desired, use the definition: $h = u + Pv$.

$$u_2 + P_2v_2 + KE_2 - u_1 - P_1v_1 - KE_1 = Q - W_s$$

Or

$$u_2 - u_1 = Q - W_s + P_1v_1 + KE_1 - P_2v_2 - KE_2$$

Calculate numerical values for all terms except $u_2 - u_1$.

$$P_2v_2 = \frac{20(144)(15)}{778} = 55.5 \text{ BTU/lb} \qquad P_1v_1 = \frac{100(144)(4.0)}{778} = 74.0 \text{ BTU/lb}$$

$$KE_2 = \frac{V^2}{2gJ} = \frac{(1000)^2}{(64.4)(778)} = 20.0 \text{ BTU/lb}$$

$$KE_1 = \frac{V^2}{2gJ} = \frac{(500)^2}{(64.4)(778)} = 5.0 \text{ BTU/lb}$$

$$W_s = \frac{195{,}000 \text{ ft-lbf}}{\text{lb}_m} \; \frac{\text{BTU}}{778 \text{ ft-lb}_f} = +\,250.6 \text{ BTU/lb}$$

Therefore,

$$u_2 - u_1 = -10 - (+250.6) + 74.0 + 5.0 - 55.5 - 20.0 = -257.1 \text{ BTU/lb decrease}$$ ●

THERMO 17

An engine developing 30 brake horsepower must be cooled by a radiator through which water is pumped from the engine. Using the following factors:

Heat transferred from engine to cooling water, expressed as horsepower, equals 40 HP.

Water temperatures in radiator:
Top water = 200°F.
Bottom water = 190°F.

Weight of 1 gallon of water = 8.33 lbs.

Calculate the required rate of flow of water in gallons per minute.

Basis of calculation: 1 minute.

$$\text{Heat loss to cooling water} = 40 \text{ HP}\left(\frac{2545 \text{ BTU}}{\text{HP-hour}}\right)\left(\frac{1 \text{ hr}}{60 \text{ min}}\right) = 1700 \text{ BTU/min}$$

Heat loss = heat gain at steady state.

Heat gain by cooling water = 1700 BTU/min = $\dot{m}C_p\Delta T$, where $\dot{m}$ is water mass flow rate in lb/min.

$$\dot{m} = \frac{1700}{(1.0)(10)} = 170 \text{ lb. water/min.}$$

$$\text{Volume flow rate} = \frac{170 \text{ lb.}}{\text{min.}} \times \frac{\text{gal.}}{8.33 \text{ lb.}} = 20.4 \text{ gal/min.}$$ ●

THERMO 18

Saturated steam is supplied to a heat exchanger. 500 gallons per minute of water are being heated from 60° to 140°F in the exchanger. Condensate is being discharged to a receiver which is vented to the atmosphere.

A. If the entering steam is supplied to the heat exchanger at a pressure of 120 psia, and is condensing at this pressure:

(1) How many pounds per hour of steam are used?

(2) What percent of condensate is lost by flashing?

B. If the pressure of the steam supplied is reduced by a pressure reducing valve so that the steam is condensing at 30 psia, then:

(3) How many pounds per hour of steam are used?

(4) What percent of condensate is lost by flashing?

C. If fuel for the boiler supplying steam to the above heat exchanger costs 40¢ per million BTU's and overall boiler efficiency is 80 percent, which of the above systems would you use? Assume condensate is returned to the boiler at 212°F and make-up water is at 60°F.

ENTHALPY

	14.7 psia	30 psia	120 psia
Saturated Liquid, h_f	180.1	218.8	312.5
Evaporation, h_{fg}	970.3	945.2	878.1
Saturated Vapor, h_g	1150.4	1164.0	1190.6

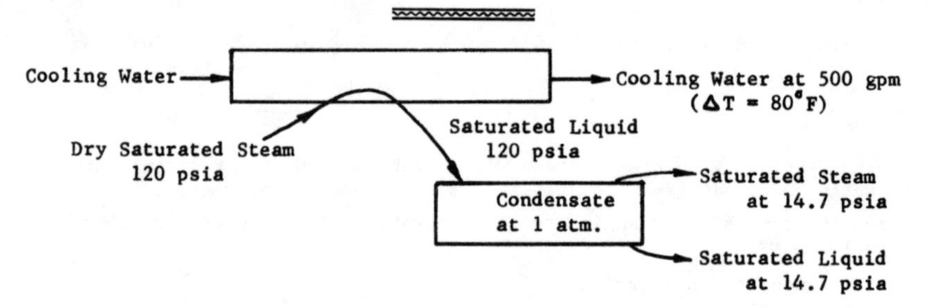

A. (1) Calculate steam flow rate (Basis: 1 hour)

Water mass flow rate, $\dot{m}_1 = \frac{500 \text{ gal}}{\text{min}} \times \frac{60 \text{ min}}{\text{hour}} \times \frac{8.33 \text{ lb.}}{\text{gal.}} = 250{,}000 \frac{\text{lb.}}{\text{hour}}$

Heat balance: heat to water = heat from condensing steam

$$\dot{m}_1 C_p (T_2 - T_1) = \dot{m}_2 h_{fg},$$ where $\dot{m}_2$ is steam mass flow rate, lb/hr.

From saturated steam table at 120 psia, $h_{fg} = 878.1$ BTU/lb.

$$\dot{m}_2 = \frac{\dot{m}_1 C_p (T_2 - T_1)}{h_{fg}} = \frac{250{,}000(1)(140 - 60)}{878.1} = 22{,}800 \frac{\text{lb.}}{\text{hour}}$$ ●

A. (2) <u>Percent condensate lost by 120 psia flashing</u> (Basis: 1 lb.)

Mass balance: X = Weight fraction flashed off.

1 lb. sat. liquid at 120 psia = (X) lb. sat. vapor at 14.7 psia
+ (1 - X) lb. sat. liquid at 14.7 psia

Heat balance:

$1(h_f \text{ at } 120 \text{ psia}) = X(h_g \text{ at } 14.7 \text{ psia}) + (1 - X)(h_f \text{ at } 14.7 \text{ psia})$

$1(312.5) = X(1150.4) + (1 - X)(180.1)$

$312.5 = 1150.4X + 180.1 - 180.1X$

$X = 0.1365$ Therefore 13.65% of condensate lost by flashing ●

B. Throttle steam to 30 psia (a constant enthalpy process)

During throttling: $h_{in} = h_{out}$

Saturated steam at 120 psia (h_g = 1190.6) becomes superheated steam at 30 psia (h_g = 1190.6).

From saturated steam table at 30 psia, h_f = 218.8 BTU/lb.

Heat liberated from steam: 1190.6 - 218.8 = 971.8 BTU/lb.

(3) <u>Calculate steam flow rate</u> (Basis: 1 hour)

Use heat balance as in (1) above: heat to water = heat from steam

$\dot{m}_1 C_p (T_2 - T_1) = \dot{m}_2 (971.8)$

$\dot{m}_2 = \frac{250{,}000(1)(140 - 60)}{971.8} = 20{,}600$ lb./hour ●

(4) <u>Percent condensate lost by 30 psia flashing</u> (Basis: 1 lb.)

Mass balance: Y = Weight fraction flashed off

1 lb. sat. liquid at 30 psia = (Y) lb. sat. vapor at 14.7 psia
+ (1 - Y) lb. sat. liquid at 14.7 psia

Heat balance:

$1(h_f \text{ at } 30 \text{ psia}) = Y(h_g \text{ at } 14.7 \text{ psia}) + (1 - Y)(h_f \text{ at } 14.7 \text{ psia})$

$1(218.8) = Y(1150.4) + (1 - Y)(180.1)$

$Y = 0.0399$ Therefore 3.99% of condensate lost by flashing ●

C. <u>Proposal Comparison</u>

	120 psia condensation and flashing	30 psia condensation and flashing
Steam rate, lb./hour	22,800.	20,600.
Steam and Condensate lost, lb/hour	3,120.	820.
Condensate returned, lb/hour	19,680.	19,780.

Superficially, it is apparent that the 30 psia case loses much less heat to the environment by flashing. Limitation on system operability is imposed by temperature on the steam side of the heat exchanger. Temperature of 120 psia saturated steam is 341°F, and superheated steam at 30 psia (h = 1190.6) is determined from superheated steam tables to be 303°F. The lower temperature requires more condenser area.

The magnitude of the heat saving is shown by an overall heat balance.
Overall heat balance on water-steam side of the boiler (Basis: 1 hour)

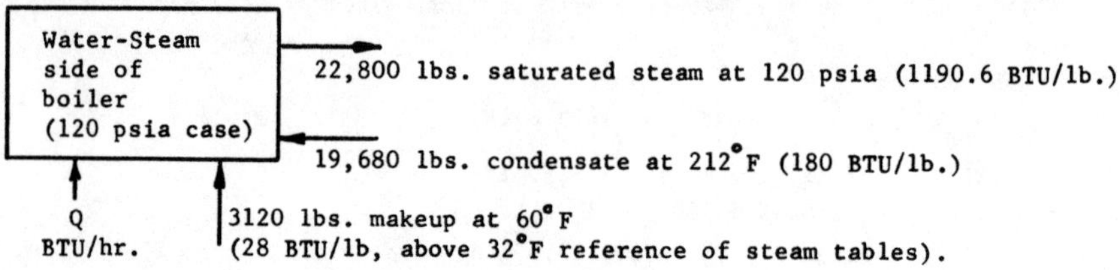

Water-Steam side of boiler (120 psia case)
22,800 lbs. saturated steam at 120 psia (1190.6 BTU/lb.)
19,680 lbs. condensate at 212°F (180 BTU/lb.)
Q BTU/hr.
3120 lbs. makeup at 60°F (28 BTU/lb, above 32°F reference of steam tables).

heat in = heat out

$$Q + 3120(28) + 19,680(180) = 22,800(1190.6)$$

$$Q + 0.09 \times 10^6 + 3.54 \times 10^6 = 27.15 \times 10^6$$

$$Q = 23.52 \times 10^6 \text{ BTU/hour}$$

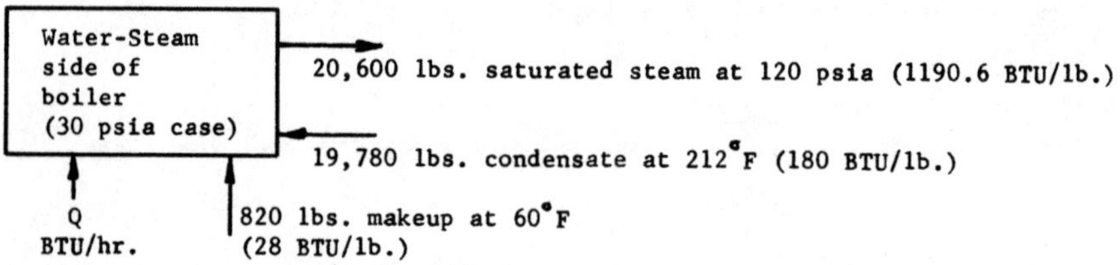

Water-Steam side of boiler (30 psia case)
20,600 lbs. saturated steam at 120 psia (1190.6 BTU/lb.)
19,780 lbs. condensate at 212°F (180 BTU/lb.)
Q BTU/hr.
820 lbs. makeup at 60°F (28 BTU/lb.)

heat in = heat out

$$Q + 820(28) + 19,780(180) = 20,600(1190.6)$$

$$Q + 0.02 \times 10^6 + 3.56 \times 10^6 = 24.55 \times 10^6$$

$$Q = 20.97 \times 10^6 \text{ BTU/hour}$$

Summary

Heat delivered to the water-steam side costs $\frac{\$0.40}{.80} = \$0.50/10^6$ BTU and the 30 psia case offers a fuel economy, but the condenser heat transfer area requirement may preclude its feasibility. A better thermal approach would be to return 120 psia sat'd liq. from condenser to boiler. ●

THERMO 19

55,000 gallons of water passes through a heat exchanger and absorbs 28,000,000 BTU's. The exit temperature is 110°F. The entrance water temperature in °F is nearest to:

(a) 49.
(b) 56.
(c) 68.
(d) 73.
(e) 82.

C_p for liquid water = 1.0 BTU/(lb.°F), density = 8.33 lb./gallon.

$$Q = mC_p\Delta T = mC_p(T_2 - T_1)$$

$$28,000,000 = (55,000 \text{ gal.})\left(\frac{8.33 \text{ lb.}}{\text{gallon}}\right)(1.0)(110 - T_1)$$

$$61.2 = 110 - T_1 \qquad T_1 = 48.8^\circ F$$ ●

Answer is (a)

THERMO 20

The mass flow rate of Freon 12 through a heat exchanger is 10 pounds/minute. Enthalpy of Freon entry is 102 BTU/lb. and of Freon exit is 26 BTU/lb. Water coolant is allowed to rise 10°F. The water flow rate in pounds/minute is:

(a) 24.
(b) 76.
(c) 83.
(d) 112.
(e) 249.

Basis of calculation: 1 minute

Heat gain by water - Heat lost by Freon = 0

$$m_1 C_p \Delta T - m_2(h_2 - h_1) = 0$$

$$m_1(1)(10) - 10(26 - 102) = 0 \qquad m_1 = \frac{760}{10} = 76. \text{ lbs./min.}$$

Answer is (b)

THERMO 21

Five gallons per minute of hot water at 180°F is produced in a flow system by injection and condensation of low pressure steam at 20 psi (gage) and 80% quality into cold water at 60°F. Calculate the steam flow rate in pounds per minute.

Calculation basis: 1 minute

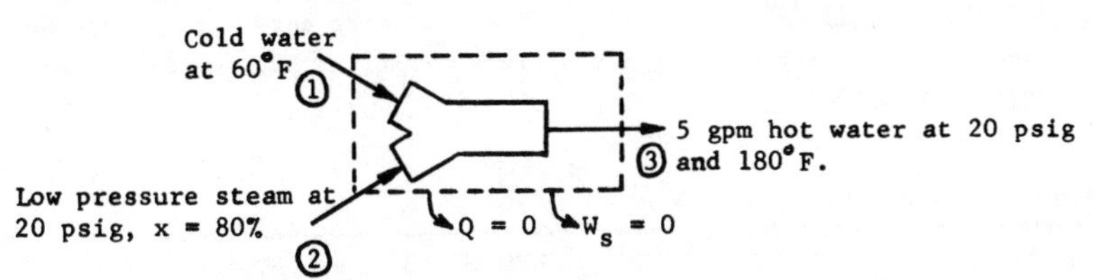

(1) First law energy balance, ignoring any PE and KE changes, for the flow system.

$$\dot{m}_3 h_3 - \dot{m}_1 h_1 - \dot{m}_2 h_2 = Q - W_s = 0$$ since no heat is lost or work done.

(2) Mass flow rate balance: $\dot{m}_1 = \dot{m}_3 - \dot{m}_2$

$$\dot{m}_3 = \frac{5 \text{ gal}}{\text{min}} \; \frac{8.33 \text{ lb.}}{\text{gal.}} = 41.67 \text{ lb./min}$$

A more precise calculation would utilize saturated steam table data at 180°F where $v_f = 0.01605 \text{ ft}^3/\text{lb.}$

$$\dot{m}_3 = \frac{5 \text{ gal}}{\text{min}} \; \frac{\text{ft}^3}{7.48 \text{ gal}} \times \frac{\text{lb.}}{0.01605 \text{ ft}^3} = 41.7 \text{ lb./min}$$

(This allows for the trivial thermal expansion of water.)

(3) Determine enthalpies from saturated steam tables:

h_1 = enthalpy of liq. water at 60°F = h_f = 28.07 BTU/lb.

h_3 = enthalpy of liq. water at 180°F = h_f = 147.91 BTU/lb.

h_2 = enthalpy of 34.7 psia (20 + 14.7) steam of 80% quality:

Saturated Steam Table Data:

psia	h_f	h_{fg}	h_g
30	218.83	945.2	1164.0
40	236.02	933.7	1169.7

Double interpolation is required: for 80% quality, and for 34.7 psia pressure.

at 80% quality, (30 psia) h = 218.83 + 0.80(945.2) = 975.0 BTU/lb.
(40 psia) h = 236.02 + 0.80(933.7) = 983.0 BTU/lb.

Linearly interpolate between 30 and 40 for 34.7 psia at 80% quality:

$$h_2 = 975.0 + \frac{4.7}{10}(983.0 - 975.0) = 978.8 \text{ BTU/lb.}$$

(4) Solve energy balance for $\dot{m}_2$, using $\dot{m}_1 = \dot{m}_3 - \dot{m}_2$

$$\dot{m}_3 h_3 = \dot{m}_1 h_1 + \dot{m}_2 h_2$$

$$(41.7)(147.91) = (41.7 - \dot{m}_2)(28.07) + \dot{m}_2(978.8)$$

$$6167.8 = 1170.5 - 28.07\dot{m}_2 + 978.8\dot{m}_2$$

$$\dot{m}_2 = \frac{4997}{950.7} = 5.26 \text{ lb/min.}$$ ●

THERMO 22

Exhaust steam from a turbine exhausts into a surface condenser at a mass flow rate of 8000 lb/hr, 2 psia and 92% quality. Cooling water enters the condenser at 74°F and leaves at the steam inlet temperature. What is the cooling water mass flow rate in lb/hr?

Saturated steam table data at 2 psia are:

T °F	h_f, BTU/lb	h_{fg}, BTU/lb	h_g, BTU/lb
126.08	93.99	1022.2	1116.2

h_1 = enthalpy of steam at 92% quality = $h_f + 0.92h_{fg}$
= 94.0 + 0.92(1022.2) = 1034.4 BTU/lb.

h_2 = enthalpy of liquid water at 126.1°F = 94.0 BTU/lb.

h_3 = enthalpy of liquid water at 74°F = 42.0 BTU/lb above reference of 32°F

Alternate: find h_f in steam tables at 74°F ignoring negligible effect of pressure on enthalpy of incompressible liquids.

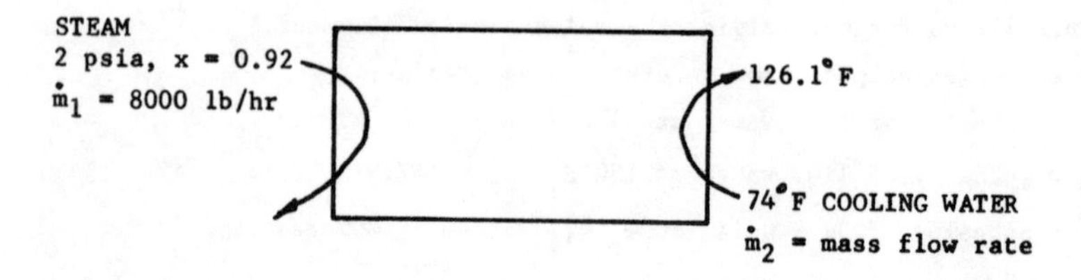

In the absence of data, assume steam condensate leaves at 126.1°F

Heat balance, basis: 1 hour

heat from steam = heat to cooling water

$\dot{m}_1(h_1 - h_2) = \dot{m}_2(h_2 - h_3)$

$8000(1034.4 - 94.0) = \dot{m}_2(94.0 - 42.0)$ $\qquad \dot{m}_2 = 144{,}700.$ lb/hr ●

THERMO 23

53.8 grams of finely divided aluminum is heated to 98.3°C and dropped into 76.2 grams of water at 18.6°C contained in a calorimeter. The final temperature of the mixture is 27.4°C. The mass of the calorimeter is 123 grams, and its specific heat may be taken as 0.092 cal $g^{-1}deg^{-1}$. The combined thermal capacity of the thermometer and metal stirrer is 6.5 cal deg^{-1}.

Assuming no heat is lost from the system, calculate the mean specific heat of aluminum for the above temperature range.

$$m_{Al}C_{Al}(T_1 - T_2) = \left[m_{cal}C_{cal} + m_{thermo}C_{thermo} + m_{H_2O}C_{H_2O}\right](T_2 - T_0)$$

$$53.8C_{Al}(98.3 - 27.4) = \left[123(0.092) + 6.5 + 76.2(1)\right](27.4 - 18.6)$$

$$3815C_{Al} = (11.31 + 6.5 + 76.2)(8.8)$$

$$C_{Al} = \frac{(94.01)(8.8)}{3815} = 0.217 \text{ cal/(gm °C)}$$ ●

THERMO 24

3000 cfm of 65°F air is required to maintain a house at 76°F.

25%wt of air entering the air conditioner is from outside at 90°F, and the remainder is recycled from inside at 76°F.

REQUIRED: Determine the rating in tons of the air conditioner.
(Assume sensible heat changes only, ignore any dehumidification.)

65°F Air Density: $\rho_{65} = \frac{P}{RT} = \frac{14.7 \times 144}{53.3 \times 525 \text{ R}} = 0.0756 \text{ lbm/ft}^3$

Total mass flow rate $\dot{m}_T = \rho Q = 0.0756(3000) = 227$ lbm/min
$= 13{,}620$ lbm/hr

$\dot{m}_{90}$ = mass flow rate 90° Air = 0.25(13,620) = 3405 lbm/hr

$\dot{m}_{76}$ = mass flow rate 76° Air = 0.75(13,620) = 10,215 lbm/hr

C_p = specific heat of Air = 0.24 BTU/lbm°F

$$Q_c = \dot{m}_{90}C_p\Delta T + \dot{m}_{76}C_p\Delta T = 3405(0.24)(90° - 65°) + 10{,}215(0.24)(76° - 65°)$$
$$= 20{,}430 + 26{,}970 = 47{,}400 \text{ BTU/hr}$$

1 ton of refrigeration = 200 BTU/min = 12,000 BTU/hr

Tons of Air Conditioning $= \frac{47{,}400}{12{,}000} = 3.95$ ●

THERMO 25

In terms of Q_H (heat from high temperature source) and Q_L (heat to low temperature sink), the net work of a Carnot cycle is:

(a) $W = Q_H - Q_L$

(b) $W = \dfrac{Q_H - Q_L}{Q_L}$

(c) $W = \dfrac{Q_L}{Q_H - Q_L}$

(d) $W = Q_L - Q_H$

(e) none of these

The work produced by a carnot cycle heat engine is $Q_H - Q_L$ ●

Answer is (a)

THERMO 26

Select the relationship defining absolute temperature in terms of Q_H and Q_L, heats transferred in a Carnot cycle, and T_H and T_L, the absolute temperatures of the reservoirs involved.

(a) $\dfrac{Q_H - Q_L}{T_H - T_L} = 1$ (b) $\dfrac{Q_H - Q_L}{Q_H} = \dfrac{T_H}{T_L}$ (c) $\dfrac{Q_H}{Q_L} = \dfrac{T_H}{T_L}$

(d) $\dfrac{Q_H}{Q_L} = \dfrac{T_L}{T_H}$ (e) $\dfrac{Q_H - Q_L}{Q_H} = \dfrac{T_H - T_L}{T_H}$

Statement (c) defines absolute thermodynamic temperature. Statement (e) is numerically equal to efficiency η of a Carnot cycle heat engine. (c) is correct, and (e) is derived therefrom. ●

THERMO 27

The maximum thermal efficiency that can be obtained in an ideal reversible heat engine operating between 1540°F and 340°F is closest to

(a) 100%
(b) 60%
(c) 78%
(d) 40%
(e) 22%

$$\eta_{thermal} = \frac{W}{Q_H} = \frac{Q_H - Q_L}{Q_H}$$

$$T_L = 340^\circ F + 460^\circ = 800^\circ R$$

$$T_H = 1540^\circ F + 460^\circ = 2000^\circ R$$

$$= 1 - \frac{Q_L}{Q_H} = 1 - \frac{T_L}{T_H}$$

$$= 1 - \frac{800}{2000} = 1 - 0.40 = 0.60 = 60\%$$ ●

Answer is (b)

THERMO 28

A 3 HP refrigerator or heat pump operates between 0°F and 100°F. The maximum theoretical heat that can be transferred from the cold reservoir is nearest to:

(a) 7,600 BTU/hr
(b) 13,000 BTU/hr
(c) 23,000 BTU/hr
(d) 35,000 BTU/hr
(e) 43,000 BTU/hr

Coefficient of Performance of a refrigerator or heat pump:

$$\text{C.O.P.} = \frac{T_L}{T_H - T_L} = \frac{460°R}{560° - 460°} = 4.6$$

$$\text{C.O.P.} = \frac{Q_L}{W} = 4.6 \qquad 1\ \text{HP} = 2545\ \text{BTU/hr}$$

$$\therefore W = 7635\ \text{BTU/hr}$$

$$\text{C.O.P.} = \frac{Q_L}{Q_H - Q_L} = \frac{Q_L}{W} = \frac{Q_L}{7635}$$

$$Q_L = 35,100\ \text{BTU/hr}$$ ●

Answer is (d)

THERMO 29

A Carnot cycle heat engine operating between 1540°F and 440°F has an efficiency of approximately

(a) 55%
(b) 45%
(c) 35%
(d) 29%
(e) 82%

For a Carnot cycle heat engine, efficiency $\eta = \frac{T_H - T_L}{T_H - 0} = \frac{2000°R - 900°R}{2000°R}$

$$= 0.55$$ ●

Answer is (a)

THERMO 30

Second law limitation on the maximum horsepower output from any power unit burning 1,000,000 BTU/hr of fuel with high and low temperature extremes of 1540°F and 40°F is:

(a) 98.
(b) 295.
(c) 1140.
(d) 3830.
(e) none of these

Basis of calculation: 1 hour

Carnot efficiency $\eta = \frac{T_H - T_L}{T_H - 0} = \frac{2000°R - 500°R}{2000°R} = 0.75 = \frac{Q_H - Q_L}{Q_H - 0} = \frac{W}{Q_H}$

$W = 0.75Q_H = 0.75 \times 10^6 = 750{,}000$ BTU/hr

$= \dfrac{750{,}000 \text{ BTU/hr}}{2545 \text{ BTU/(HP hr)}} = 295.$ HP ●

Answer is (b)

THERMO 31

A heat pump installation is used to warm a stream of circulating air in a residence to 110°F. An outside ambient of 35°F is considered the available heat source. A heating requirement of 90,000 BTU/hr is necessary to maintain a comfortable temperature in the living space.

(a) Determine the absolute minimum electrical requirement in kilowatts and in HP to operate a heat pump delivering 90,000 BTU/hr under these conditions.

(b) Determine the electrical requirement in Kw to accomplish the same heat by resistance heating.

A Carnot reversed cycle would provide the most efficient heat pump possible, although this is only approached in practice with Rankine vapor compression cycles.

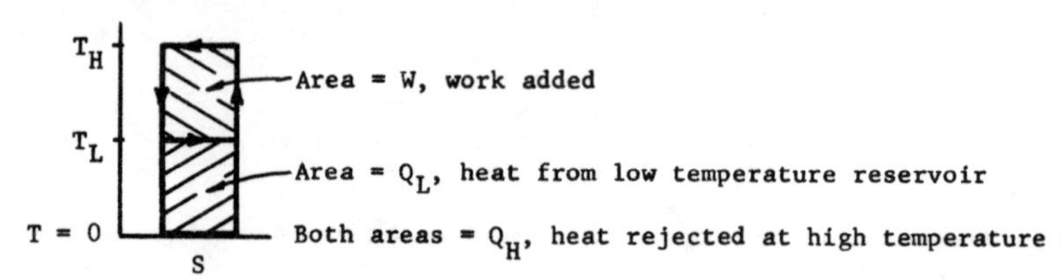

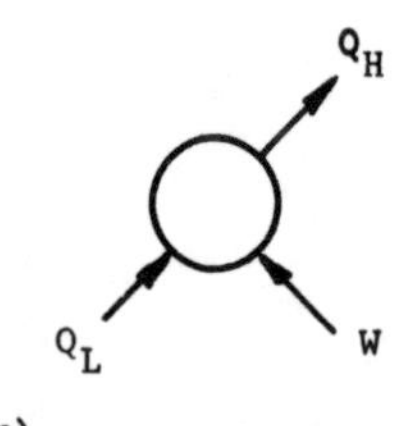

For a heat pump: $Q_H = Q_L + W$ as shown in the T-S diagram

Coefficient of Performance:

$$\text{C.O.P.} = \frac{Q_L}{W} = \frac{T_L}{T_H - T_L}$$

Absolute temperatures are required.

110°F = 570°R, 35°F = 495°R.

(a)

$$\text{C.O.P.} = \frac{Q_L}{W} = \frac{495}{570 - 495} = 6.60$$

$Q_H = 90{,}000 = Q_L + W = 7.60\ W$

$$W = 11{,}840 \text{ BTU/hr} = \frac{11{,}840 \text{ BTU/hr}}{3413 \text{ BTU/Kwh}} = 3.47 \text{ Kw} \bullet$$

$$\frac{3.47 \text{ Kw}}{0.746 \text{ Kw/HP}} = 4.65 \text{ HP} \bullet$$

(b)

Resistance heating is pure energy dissipation

$$\frac{90{,}000}{3{,}413} = 26.4 \text{ Kw} \bullet$$

THERMO 32

In any non-quasistatic thermodynamic process, the overall entropy of an isolated system will:

(a) Increase and then decrease
(b) Decrease and then increase
(c) Stay the same
(d) Increase only
(e) Decrease only

quasistatic: infinitely slow, lossless, hypothetical, by differential increments

The overall entropy will increase for an isolated system, or for the system plus surroundings. ●

Answer is (d)

THERMO 33

Entropy is the measure of

(a) the change in enthalpy of a system
(b) the internal energy of a gas
(c) the heat capacity of a substance
(d) randomness or disorder
(e) the total heat content of a system

Answer is (d) ●

THERMO 34

A quantity of working fluid undergoes three sequential irreversible processes:

(1) Receives heat from high temperature reservoir (at high ΔT)
Reservoir entropy decreases 0.240 BTU/°R and
Fluid entropy increases 0.245 BTU/°R.

(2) Does work in an adiabatic turbine.
Fluid entropy increases 0.013 BTU/°R.

(3) Discharges heat to low temperature reservoir (at high ΔT)
Fluid entropy decreases 0.258 BTU/°R.
Reservoir entropy increases 0.263 BTU/°R.

Entropy change for the universe is BTU/°R is:

(a) indeterminate
(b) -0.023
(c) +0.010
(d) +0.023
(e) none of these

Net increase for the universe equals + .023, the difference between high temperature reservoir loss and low temperature gain. This is a consequence of the irreversibilities because the heat transfer is not at $\Delta T = 0$, and there is lost work in the turbine. ●

Answer is (d)

THERMO 35

If heat transfers of the previous problem were reversible, and if the turbine operated isentropically, what is the entropy gained or lost by the low temperature reservoir?

(a) indeterminate
(b) 0
(c) +0.240 BTU/°R
(d) +0.250 BTU/°R
(e) none of these

Reservoirs are considered infinite and no temperature rise occurs. Since ΔS for reversible heat transfer = 0, and the turbine is isentropic ($\Delta S = 0$), entropy loss by the high temperature reservoir matches the entropy gain by the low temperature reservoir = +0.240 BTU/°R. ●

Answer is (c)

THERMO 36

A Carnot cycle heat engine operates between 1540°F and 40°F and rejects 250. BTU/lb to the low temperature reservoir or heat sink. Maximum entropy change in BTU/lb°R of the working fluid for any process in the cycle is:

(a) 0
(b) 0.25
(c) 0.50
(d) 1.0
(e) 4.0

The T-S diagram for a Carnot heat engine shows its relation between heat temperature and entropy.

$Q_{rev} = T\Delta S$

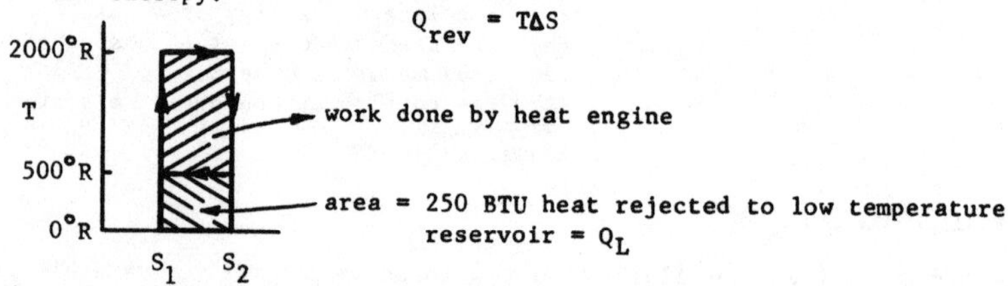

Both areas = heat intake from high temperature reservoir = Q_H.

Since $250 = 500 \cdot \Delta S$ $\therefore \Delta S = 0.50$ ●

Answer is (c)

THERMO 37

For spontaneously occurring natural processes in an isolated system, which expression best expresses dS?

(a) $dS = \frac{dQ}{T}$
(b) $dS = 0$
(c) $dS > 0$
(d) $dS < 0$
(e) $dS = C_p\frac{dt}{T} - R\frac{dp}{P}$

(a) $dS = \frac{dQ_{rev}}{T}$ only. The reversible requirement is necessary to generate the exact height vs. rectangular area equivalence on the Carnot cycle T-S diagram.
(b) Only a reversible adiabatic process is isentropic by definition.
(c) All naturally occurring spontaneous processes are irreversible and result in an entropy increase. ●
(d) An energy input from surroundings is required to reduce entropy.
(e) This is an expression for entropy change in an ideal gas.

Answer is (c)

THERMO 38

Which of the following statements about entropy is false?

(a) Entropy of a mixture is greater than that of its components under the same conditions.

(b) An irreversible process increases entropy of the universe.

(c) Entropy has the units of heat capacity.

(d) Net entropy change in any closed cycle is zero.

(e) Entropy of a crystal at $0^\circ F$ is zero.

All are true except (e). The entropy of a perfect crystal at absolute zero ($0^\circ K$ or $0^\circ R$) is zero. This is the 3rd law of thermodynamics. There is presumably no randomness existent at this temperature in a crystal without flaws, impurities or dislocations.

Answer is (e) ●

THERMO 39

Work or energy can be a function of all of the following except:

(a) force and distance

(b) torque and angular rotation

(c) power and time

(d) force and time

(e) temperature and entropy

(a) $dW = FdS$ for translation

(b) $dW = Td\theta$ for rotation

(c) $\frac{dW}{dt} = \text{power}$ $\therefore dW = (\text{Power})dt$

(d) $Fdt = d(mV)$, the impulse momentum equation. Impulse, Fdt, is not an energy term.

(e) $dQ_{rev} = TdS$. Heat is a form of energy

All except (d) are correct ●

THERMO 40

Energy changes are represented by all except which one of the following?

(a) $-\int VdP$

(b) $TdS - PdV$

(c) $TdS + VdP$

(d) $\frac{dQ_{rev}}{T}$

(e) $mC_p dT$

(a) is shaft work.

(b) $dU = TdS - PdV = dQ_{rev} - W$. dU is an internal energy change.

(c) $dH = TdS + VdP$. dH is an enthalpy change.

(d) $\frac{dQ_{rev}}{T} = dS$. dS is change in the thermodynamic state function, entropy. It is not an energy change.

(e) $dH = mC_p dT$ for an ideal gas.

Therefore (d) represents an entropy or state change, all others represent energy changes. ●

Answer is (d)

THERMO 41

A high velocity flow of gas at 800 ft/sec possesses kinetic energy nearest to which of the following?

(a) 1.03 BTU/lb
(b) 4.10 BTU/lb
(c) 9.95 BTU/lb
(d) 12.8 BTU/lb
(e) 41.0 BTU/lb

Basis: 1 lb_m of fluid flowing

$KE = \frac{V^2}{2g_c}$ in ft-lb_f, where V is in ft/sec, and $g_c = 32.17$

Use J = 778 ft-lb_f/BTU to convert to BTU.

$$KE = \frac{800^2}{2(32.17)(778)} = 12.8 \text{ BTU/lb.}$$ ●

Answer is (d)

THERMO 42

(U + PV) is a quantity called

(a) flow energy
(b) shaft work
(c) entropy
(d) enthalpy
(e) internal energy

Flow energy is PV. Shaft work, W_s, is $-\int VdP$. Entropy is S.

Internal energy is U. Enthalpy H is defined as U + PV, the sum of internal energy plus flow energy. ●

Answer is (d)

THERMO 43

In flow process, neglecting KE and PE changes, $-\int VdP$ represents which item below?

(a) heat transfer
(b) shaft work
(c) closed system work
(d) flow energy
(e) enthalpy change

Shaft work is work or mechanical energy crossing the fixed size boundary (control volume) of a flow (open) system. Shaft work W_s is defined, in the absence of PE and KE changes, by $dH = TdS + VdP$, where $TdS = dQ_{rev}$ and $-VdP$ is dW_s. In integrated form $\Delta H = \int TdS + \int VdP = Q_{rev} - W_s$, where W_s is represented by $-\int VdP$. Closed system work W is defined by $dU = TdS - PdV$, or $\Delta U = \int TdS - \int PdV = Q_{rev} - W$. Thus closed system work is $+\int PdV$.

Flow energy is the PV term, and enthalpy change is ΔH.

Answer is (b) ●

THERMO 44

Power may be expressed in units of

(a) ft-lbs
(b) BTU/hr
(c) HP-hours
(d) Kw-hours
(e) BTU

Ft-lb, HP-hours, BTU, and Kwh are usual mechanical, thermal and electrical energy terms. Power = energy/time. Usual power units are ft-lb/sec, HP, BTU/hr and Kw.

Answer is (b) ●

THERMO 45

Given the following data:

Electricity cost	\$0.015/Kwh
Natural gas cost	\$0.065/100 cu ft
Heat content of gas	1050 BTU/cu ft

Required: Compute how many times as expensive it would be to heat a house by electricity than to heat the same house by gas, if the electric heat is assumed to be 100% efficient and the gas is 60% efficient.

We will need to convert foot-pounds to BTU or vice versa

1 BTU = 778 foot-pounds

Heating Cost:

$$\text{Natural Gas} = \frac{\$0.065}{0.60 \times 1050 \times 100} = \$1.03 \times 10^{-6}/\text{BTU}$$

$$\text{Electricity} = \frac{\$0.015 \times 0.746\text{Kw/HP} \times 1\ \text{HP}/33{,}000\ \text{ft-lb/min} \times 1/60 \times 778}{1.0}$$

$$= \$4.40 \times 10^{-6}/\text{BTU}$$

$$\text{Ratio of } \frac{\text{Cost of electrical heating}}{\text{cost of gas heating}} = \frac{\$4.40 \times 10^{-6}/\text{BTU}}{\$1.03 \times 10^{-6}/\text{BTU}} = 4.27 \quad ●$$

THERMO 46

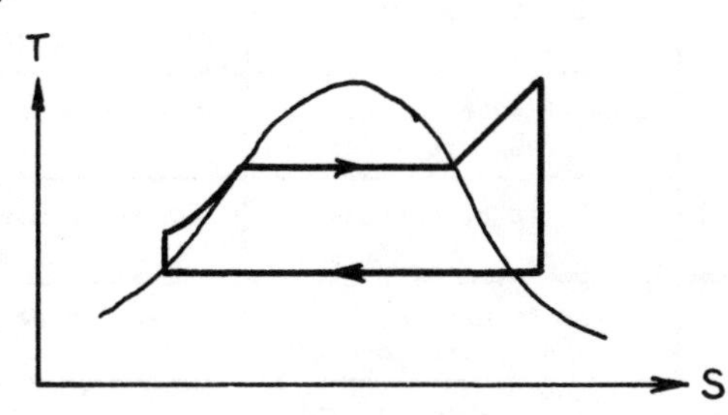

The above temperature-entropy diagram represents:

(a) Rankine cycle with superheated vapor
(b) Carnot cycle
(c) Diesel cycle
(d) Refrigeration cycle
(e) Adiabatic process

Answer is (a) ●

THERMO 47

A Carnot heat engine cycle is represented on the T-S and P-V diagrams below:

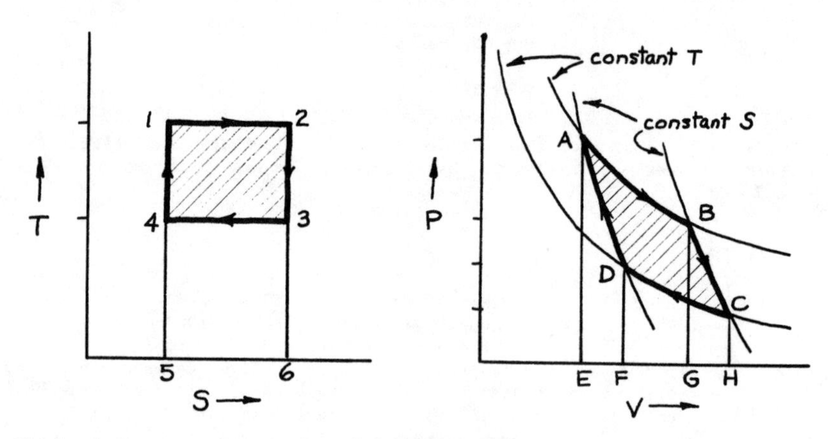

Which of the several areas bounded by numbers or letters represents the amount of heat rejected by the fluid during one cycle?

(a) area 1-2-6-5
(b) area B-C-H-G
(c) area 3-4-5-6
(d) area D-A-E-F
(e) area C-D-F-H

The table below gives the significance of each area of the diagrams:

PROCESS	T-S Diagram Area Representing Heat	P-V Diagram Area Representing Work
isothermal expansion 1-2 and A-B	1-2-6-5 = heat in from high temp. reservoir	A-B-G-E = work done by fluid
isentropic expansion 2-3 and B-C	2-3-6 = 0 heat transfer	B-C-H-G = work done by fluid
isothermal compression 3-4 and C-D	3-4-5-6 = heat out to low temp. reservoir	C-D-F-H = work done on fluid
isentropic compression 4-1 and D-A	4-1-5 = 0 heat transfer	D-A-E-F = work done on fluid
net result of process	1-2-3-4 = net heat converted to work	A-B-C-D = net work done by process

Answer is (c) ●

THERMO 48

Which of the following thermodynamic cycles is the most efficient?

(a) Carnot
(b) Brayton
(c) Otto
(d) Diesel
(e) Rankine

Answer is (a)

THERMO 49

A Carnot engine operating between 70°F and 2000°F is modified solely by raising the high temperature by 150°F and raising the low temperature by 100°F. Which of the following statements is false?

(a) thermodynamic efficiency is increased.
(b) more work is done during the isothermal expansion.
(c) more work is done during the isentropic compression.
(d) more work is done during the reversible adiabatic expansion.
(e) more work is done during the isothermal compression.

Carnot cycle efficiency is originally

$$\eta = \frac{T_H - T_L}{T_H - 0} = \frac{2460°R - 530°R}{2460°R} = 0.785$$

After the change

$$\eta = \frac{2610 - 630}{2610} = 0.76$$ Efficiency is therefore reduced.

On the T-S and P-V diagrams below the original cycle is shown as ABCD, the modified cycle as A'B'C'D'.

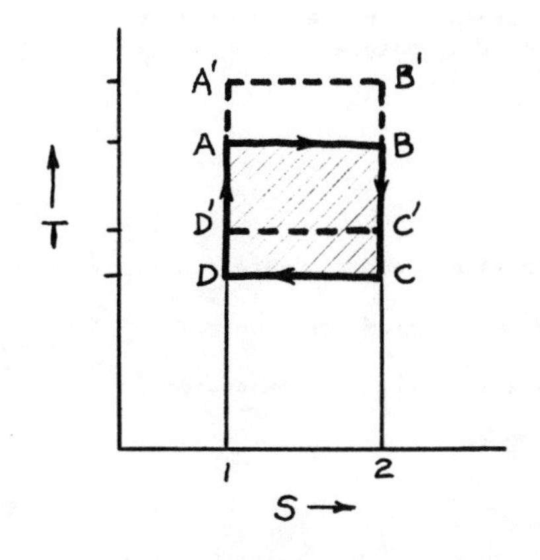

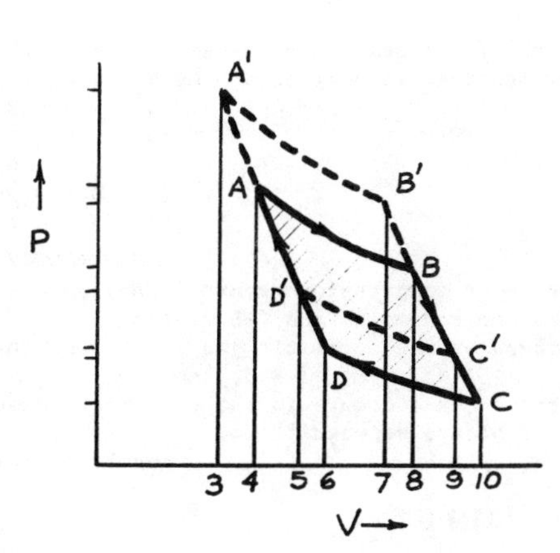

Compare work done during isothermal expansion (A to B, vs. A' to B')
 Original: area A-B-8-4
 Modified: area A'-B'-7-3 is larger

Compare work done during isentropic compression (D to A, vs. D' to A')
 Original: area D-A-4-6
 Modified: area D'-A'-3-5 is larger

Compare work during reversible (isentropic) expansion (B to C, vs. B' to C')
 Original: area B-C-10-8
 Modified: area B'-C'-9-7 is larger

Compare work during isothermal compression (C to D, vs. C' to D')
 Original: area C-D-6-10
 Modified: area C'-D'-5-9 is larger

Statements (b), (c), (d), and (e) are correct.

The false answer is (a) ●

THERMO 50

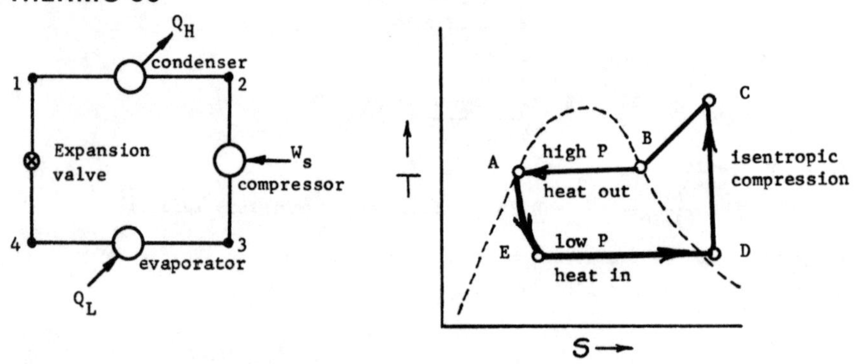

In the ideal heat pump system outlined, the expansion valve 4-1 performs the process that is located on the T-S diagram between points

(a) A and B
(b) B and C
(c) C and D
(d) D and E
(e) E and A

The vapor compression reversed Rankine cycle is conducted counterclockwise on both the schematic and T-S diagrams.
Numbers on the schematic and letters on the T-S diagram are related:
 1 = A, 2 = B, 3 = D, and 4 = E.
Process C-B-A occurs in the condenser between 2 and 1. The expansion process A - E occurs between 1 - 4.

Answer is (e) ●

THERMO 51

Which air-standard power cycle do the P-V and T-S diagrams on the next page represent?

(a) Otto cycle
(b) Reheat cycle
(c) Carnot cycle
(d) Rankine cycle
(e) Brayton cycle

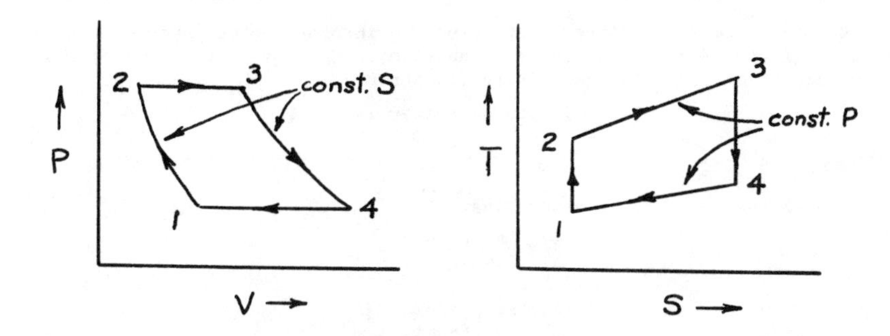

The Brayton cycle is applied to the simple open-cycle gas turbine wherein intake air is compressed (1-2), combustion supplies thermal energy (2-3), and combustion products expand and drive the turbine (3-4), and exhaust at 4. ●

Answer is (e)

THERMO 52

Equilibrium conditions exist in all <u>except</u> which of the following:

(a) in reversible processes.
(b) in processes where driving forces are infinitesimals.
(c) along ideal frictionless, nondissipative paths where forward and reverse processes occur at equal rates.
(d) in a steady state flow process.
(e) where nothing can occur without an effect on the system's surroundings.

All except (d) demonstrate equilibrium. Steady state or time invarient flow processes may be conducted far from equilibrium conditions. ●

The <u>false</u> answer is (d)

THERMO 53

Data in the table below describe two states of a working fluid that exist at two locations in a piece of hardware.

	P psia	v ft^3/lb	T °F	h BTU/lb	s BTU/(lb°F)
State 1	25.	0.011	20.	19.2	0.0424
State 2	125.	0.823	180.	203.7	0.3649

Which of the following statements about the path from state 1 to 2 is <u>false</u>?

(a) the path results in an expansion.
(b) the path determines the amount of work done.
(c) the path is indeterminate from these data.
(d) the path requires that energy be added in the process.
(e) the path is reversible adiabatic.

The large volume and entropy changes indicate change from a condensed to vapor phase. Temperature, pressure, and enthalpy increases require an energy input.

The path from 1 to 2 is indeterminate since no information on intermediate states is given. Work is always path dependent. Entropy increase means the process cannot be reversible adiabatic (isentropic).

The <u>false</u> answer is (e) ●

THERMO 54

Name the process that has no heat transfer.

(a) Isentropic
(b) Isothermal
(c) Quasistatic
(d) Reversible
(e) Polytropic

An <u>isentropic</u> process is reversible adiabatic. An <u>adiabatic</u> process has no heat exchange with surroundings. An <u>isothermal</u> process is conducted at constant temperature. A <u>quasistatic</u> (almost static) process departs only infinitesimally from an equilibrium state. A <u>reversible</u> process can have its initial state restored without any change (energy gain or loss) taking place in the surroundings. A <u>polytropic</u> process is conducted with changes in temperature, pressure, volume, and entropy; it follows the relationship PV^n = constant, where $n \neq \frac{C_p}{C_v}$.

Answer is (a) ●

THERMO 55

In an open or flow process (across a fixed control volume) in the absence of PE and KE changes, which of the following represents the shaft work done during a polytropic process?

(a) $W_s = \frac{k}{1-k}(P_2V_2 - P_1V_1) = \frac{k}{1-k}mR(T_2 - T_1) = \frac{k}{1-k}mRT_1\left[\left(\frac{P_2}{P_1}\right)^{\frac{k-1}{k}} - 1\right]$

(b) $W_s = P_1V_1\ln(\frac{P_1}{P_2}) = P_1V_1\ln(\frac{V_2}{V_1}) = mRT\ln(\frac{V_2}{V_1}) = mRT\ln(\frac{P_1}{P_2})$

(c) $W_s = \frac{n}{1-n}(P_2V_2 - P_1V_1) = \frac{n}{1-n}mR(T_2 - T_1) = \frac{n}{1-n}mRT_1\left[\left(\frac{P_2}{P_1}\right)^{\frac{n-1}{n}} - 1\right]$

(d) $W_s = 0$

(e) $W_s = -V(P_2 - P_1)$

Shaft work, W_s, for a flow process = $-\int VdP$. The above equations are valid for ideal gases in the processes listed below:

(a) for constant entropy (isentropic, or reversible adiabatic) process.
(b) for isothermal process (at constant T, U and H are also constant for ideal gases).
(c) for polytropic process (where P, V, T, and S all change).
(d) for constant pressure (isobaric) process.
(e) for constant volume (isochoric) process.

The equations may be modified by manipulation with the ideal gas law $PV = mRT$, $P_1V_1 = mRT_1$, $P_2V_2 = mRT_2$ in all cases. The relations that follow also may be substituted, using n or k as appropriate, for polytropic and isentropic processes:

$$PV^n = P_1V_1^{\,n} = P_2V_2^{\,n} = \text{constant}, \qquad \left(\frac{P_2}{P_1}\right)^{\frac{n-1}{n}} = \frac{T_2}{T_1} = \left(\frac{V_1}{V_2}\right)^{n-1}$$

and $\dfrac{P_2}{P_1} = \left(\dfrac{V_1}{V_2}\right)^n$. The correct answer to the question is (c). All the rest are valid when applied to the proper <u>flow</u> process. ●

THERMO 56

In a closed system (with a moving boundary) which of the following represents work done during an isothermal process?

(a) $W = P(V_2 - V_1)$

(b) $W = 0$

(c) $W = P_1V_1 \ln(\frac{P_1}{P_2}) = P_1V_1 \ln(\frac{V_2}{V_1}) = mRT \ln(\frac{V_2}{V_1}) = mRT \ln(\frac{P_1}{P_2})$

(d) $W = \dfrac{P_2V_2 - P_1V_1}{1 - k} = \dfrac{mR(T_2 - T_1)}{1 - k}$

(e) $W = \dfrac{P_2V_2 - P_1V_1}{1 - n} = \dfrac{mR(T_2 - T_1)}{1 - n}$

Work for a closed system (piston-cylinder type, non-repetitious) $W = \int PdV$.

The above equations are valid for ideal gases in the processes listed below:

(a) constant pressure
(b) constant volume
(c) isothermal process
(d) isentropic process
(e) polytropic process

Note that all modifications by manipulation noted in the previous question may be applied to this set of equations.

The Answer to the question is (c) ●

THERMO 57

Work of a polytropic (n = 1.21) compression of air ($\frac{C_p}{C_v} = 1.40$) in a system with moving boundary from $P_1 = 15.$ psia, $V_1 = 1.0\ ft^3$ to $P_2 = 150.$ psia, $V_2 = 0.15\ ft^3$ is:

(a) 35.5 ft.lb.
(b) 324 ft.lb.
(c) 1080 ft.lb.
(d) 2700 ft.lb.
(e) 5150 ft.lb.

Work of a closed system (moving boundary) polytropic process for an ideal gas is:

$$W = \frac{P_2V_2 - P_1V_1}{1 - n} = \frac{[150(0.15) - 15(1.0)]\,144}{1 - 1.21} = -5143.\ \text{ft.lb.}$$

●, work is done on gas.

Answer is (e)

THERMO 58

Isentropic compression of 1 ft^3 of air, $\frac{C_p}{C_v} = 1.40$, at 20 psia to a pressure of 100 psia gives a final volume of:

(a) 0.16 ft^3
(b) 0.20 ft^3
(c) 0.32 ft^3
(d) 0.40 ft^3
(e) 0.56 ft^3

An isentropic process for an ideal gas follows the path:

$$P\,V^k = P_1 V_1^{\,k} = P_2 V_2^{\,k} = \text{constant, where } k = \frac{C_p}{C_v}$$

$$20(1)^{1.4} = 100(V_2)^{1.4} \qquad V_2^{\,1.4} = 0.20 \qquad \therefore V_2 = 0.317\ ft^3$$

Answer is (c)

THERMO 59

Steam enters a turbine at 300 psia, 700°F and exhausts to a low pressure process steam line at 30 psia. The turbine delivers 1000 Kw of power at an engine efficiency of 72%. Calculate the mass flow rate of steam in lb/hr.

Engine efficiency is the ratio of output to the available work corresponding to an isentropic expansion.

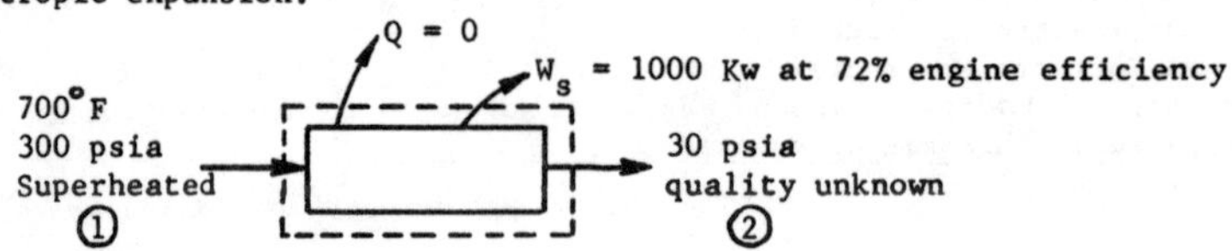

1 Kwh = 3413 BTU

Calculation basis: 1 hour

1000 Kwh at 72% efficiency is equivalent to 1389 Kwh, or

$W_s = 4.74 \times 10^6$ BTU from an isentropic expansion.

Incoming steam properties from superheated tables:

$h_1 = 1368.3$ BTU/lb $\qquad s_1 = 1.6751$ BTU/(lb°F)

Exhaust steam for isentropic expansion has $s_2 = 1.6751$ BTU/(lb°F) at 30 psia.

Consult saturated steam tables at 30 psia finding:

T, °F	h_f	h_{fg}	h_g	s_f	s_{fg}	s_g
250.34	218.83	945.2	1164.0	0.3680	1.3312	1.6992

Linearly interpolate entropy to 1.6751 and determine enthalpy on the basis of equal quality.

$s_2 - s_f = 1.6751 - 0.3680 = 1.2891$

s_{fg} for 100% quality $= 1.3312$ $\qquad X = \frac{1.2891}{1.3312} = 0.9684$

h_{fg} for 100% quality $= 945.2$

h_2 at .9684 quality $= 218.83 + 0.9684(945.2) = 1134.2$ BTU/lb.

Let $\dot{m}_1$ be steam flow rate in lb/hr.

$$\dot{m}_1(h_1 - h_2) = W_s$$

$$\dot{m}_1(1368.3 - 1134.2) = 4.74 \times 10^6 \text{ BTU}$$

$$\dot{m} = \frac{4.74 \times 10^6}{234.1} = 20{,}200 \text{ lb/hr.}$$ ●

THERMO 60

An ideal gas at a pressure of 500 psia and a temperature of 75°F is contained in a cylinder with a volume of 700 cubic feet. A certain amount of the gas is released so that the pressure in the cylinder drops to 250 psia. The expansion of the gas is isentropic. The heat capacity ratio is 1.40 and the gas constant is 53.3 ft lb_f/lb_m°R.

What is the weight of the gas remaining in the cylinder?

Given: $k = \frac{C_p}{C_v} = 1.40$ $\qquad R = 53.3 \frac{\text{ft } lb_f}{lb_m \text{ °R}}$

$P_1 = 500$ psia $\qquad P_2 = 250$ psia

$V_1 = 700$ cu ft $\qquad V_2 = 700$ cu ft

$T_1 = 75°F + 460 = 535°R$ $\qquad T_2 = ?$

$w_2 = ?$

Basis:

Ideal gas law $P_1V_1 = wRT_1$ and the basic equations for reversible adiabatic (isentropic) expansion

$$\frac{T_2}{T_1} = \left(\frac{V_1}{V_2}\right)^{k-1} \quad \text{and} \quad \frac{T_2}{T_1} = \left(\frac{P_2}{P_1}\right)^{\frac{k-1}{k}}$$

The gas remaining in the tank cools as it expands

$$T_2 = T_1\left(\frac{P_2}{P_1}\right)^{\frac{k-1}{k}} = 535\left(\frac{250}{500}\right)^{\frac{1.4-1}{1.4}} = 535\left(\frac{1}{2}\right)^{0.2857} = 439°R$$

$P_2V_2 = w_2RT_2$ $\qquad$ Check dimensions to use the proper conversion factor.

$$250 \frac{lb_f}{in^2} \times \frac{144 \text{ in}^2}{ft^2} \times 700 \text{ ft}^3 = w_2 \text{ lb}_m \frac{53.3 \text{ ft } lb_f}{lb_m \text{ °R}} \times 439°R$$

$$w_2 = \frac{(250)(144)(700)}{(53.3)(439)} = 1077 \text{ lb}_m$$ ●

THERMO 61

Determine the theoretical horsepower required for the isothermal compression of 800 ft^3/min of air from 14.7 to 120 psia.

At this volume flow rate, the process is a flow process. Work of isothermal compression of an ideal gas is numerically the same in a steady flow process as in a closed system.

$PV = \text{constant} = P_1V_1 = P_2V_2 = RT$

closed system: $W = \int_{V_1}^{V_2} PdV = P_1V_1 \ln\frac{V_2}{V_1} = P_1V_1 \ln\frac{P_1}{P_2} = mRT \ln\frac{V_2}{V_1} = mRT \ln\frac{P_1}{P_2}$

flow system: $W_s = -\int_{P_1}^{P_2} VdP = -P_1V_1 \ln\frac{P_2}{P_1} = P_1V_1 \ln\frac{P_1}{P_2} = P_1V_1 \ln\frac{V_2}{V_1}$

$= mRT \ln\frac{V_2}{V_1} = mRT \ln\frac{P_1}{P_2}$

Basis of calculation: 1 minute

$W_s = P_1V_1 \ln\frac{P_1}{P_2} = 14.7(144)(800) \ln\left(\frac{14.7}{120}\right) = -3{,}560{,}000$ ft.lb/min, work in

$W_s = \frac{3{,}560{,}000 \text{ ft.lb/min}}{33{,}000 \text{ ft.lb/HP-min}} = 108.$ HP ●

THERMO 62

Enthalpy of an ideal gas is a function only of:

(a) internal energy
(b) entropy
(c) the product of pressure and specific volume
(d) pressure
(e) temperature

$H = U + PV$, $PV = RT$. Enthalpy and internal energy of ideal gases are a function only of temperature. $dh = C_p dT$ and $du = C_v dT$.

Answer is (e) ●

THERMO 63

Which of the following statements is false concerning the deviations of real gases from ideal gas behavior?

(a) Molecular attraction interactions are compensated for in the ideal gas law.

(b) Deviations from ideal gas behavior are large near the saturation curve.

(c) Deviations from ideal gas behavior become significant at pressures above the critical point.

(d) Molecular volume becomes significant as specific volume is decreased.

(e) Compressibility factor Z is used to modify the ideal gas equation of state to fit real gas behavior.

All statements except (a) are true. The ideal gas law does not consider volume of the molecules or any interaction other than elastic collision. ●

The <u>false</u> statement is (a)

THERMO 64

There are 3 lbs. of air in a rigid container at 25 psia and 100°F. Given the gas constant for air is 53.35
(a) Determine the volume of the container
(b) If the temperature is raised to 180°F, what is the resulting absolute pressure?

$PV = mRT$

$$\frac{lb}{ft^2} \cdot ft^3 = lb_m \cdot \frac{ft - lb}{lb_m \ ^\circ R} \cdot {}^\circ R$$

$P = 25 \text{ psia} \times 144 = 3600 \text{ lb/ft}^2$

$R = 53.35 \text{ ft-lb/lb}_m \cdot {}^\circ R$

$T_1 = 100^\circ F + 460 = 560^\circ R$

$T_2 = 180^\circ F + 460 = 640^\circ R$

(a) $PV = mRT$

$3600(V) = 3(53.35)(560)$ $\quad V = \dfrac{3(53.35)(560)}{3600} = 24.9 \text{ ft}^3$ ●

(b) $\dfrac{P_1V_1}{T_1} = \dfrac{P_2V_2}{T_2}$ For constant volume

$$\frac{P_1}{T_1} = \frac{P_2}{T_2}$$

$$P_2 = \frac{P_1T_2}{T_1} = \frac{25(640)}{560} = 28.6 \text{ psia}$$ ●

THERMO 65

A mixture at 14.7 psia and 68°F that is 30% weight CO_2 (m wt = 44) and 70% weight N_2 (m wt = 28) has a partial pressure of CO_2 in psia that is nearest to:

(a) 2.14
(b) 3.15
(c) 6.83
(d) 7.86
(e) 11.55

Basis of calculation: 1 lb. mixed gases

(1) Calculate weight of each component, number of moles of each present.

(2) Compute mole fraction of each, and proportion total pressure according to mole fraction.

Component	Weight, lb.	Number of lb. moles	Mole Fraction	Partial Pressure psia
CO_2	0.30	$\frac{0.30}{44} = .00683$	$\frac{.00683}{.03183} = 0.214$	3.15
N_2	0.70	$\frac{0.70}{28} = 0.0250$	$\frac{0.0250}{0.03183} = 0.786$	11.55
Total	1.00	0.03183	1.000	14.70

Since mole fraction of a gas = volume fraction,
Composition of the mixture is 21.4% vol. CO_2 and 78.6% vol. N_2.

From the table, the correct partial pressure of CO_2 is 3.15 psia. ●

Answer is (b)

THERMO 66

Given that molar C_p of CO_2 is 8.92 BTU/(lb.mole°R), and molar C_p of N_2 is 6.95 BTU/(lb.mole°R), the calculated C_p per pound of mixture containing 25% vol. CO_2 and 75% vol. N_2 is:

(a) 0.23 BTU/(lb.°R)
(b) 2.23 BTU/(lb.°R)
(c) 5.21 BTU/(lb.°R)
(d) 7.44 BTU/(lb.°R)
(e) none of these

Note that theoretical molar heat capacities of ideal gases at zero pressure are for monotomic: $C_v = 3$, $C_p = 5$. BTU/(lb.mole°R)

diatomic: $C_v = 5$, $C_p = 7$.

triatomic: $C_v = 7$, $C_p = 9$.

Basis of calculation: 1 lb. mole of mixture

Component	Mol. Wt.	% vol.	Mole fraction	Weight of each component	Molar C_p contribution
CO_2	44	25.	0.25	11.0	0.25(8.92) = 2.23
N_2	28	75.	0.75	21.0	0.75(6.95) = 5.21
Total		100.	1.00	32.0	7.44
				Average Mol. Weight	

One mole of mixture weighs 32.0 lb. and has a Molar C_p of 7.44 BTU/(lb.mole°R)

The mixture has a specific C_p of $\frac{7.44}{32.0}$ = 0.232 BTU/(lb.°R) ●

Answer is (a)

THERMO 67

Dry air has an average molecular weight of 28.9, consisting of 21 mole-% O_2, 78 mole-% N_2 and 1 mole-% (Argon, traces of CO_2). Its calculated wt.% O_2 is nearest to:

(a) 21.0
(b) 22.4
(c) 23.2
(d) 24.6
(e) 28.0

Basis of calculation: 1 lb. mole dry air

Component	Mol. Wt.	Mole Fraction	Weight, lb.	% Weight
O_2	32.0	0.21	6.72	23.2
N_2	28.0	0.78	21.80	75.4
Ar	40.0	0.01	0.40	1.4
Total		1.00	28.92	100.0

Answer is (c) ●

THERMO 68

All of the following statements about wet bulb temperature are true EXCEPT:

(a) wet bulb temperature equals adiabatic saturation temperature
(b) wet bulb temperature lies numerically between dewpoint and dry bulb temperatures for unsaturated systems
(c) wet bulb temperature equals both dry bulb and dewpoint temperature at 100% relative humidity
(d) wet bulb temperature is the only temperature necessary to determine grains of water per lb. of dry air
(e) wet bulb temperature is the lowest temperature attainable by evaporative cooling at ambient pressure

Wet bulb temperature, as commonly measured, and adiabatic saturation temperature are equal for the air-water system. For unsaturated systems, wet bulb temperature is approximately equal to dewpoint plus 1/3 of the difference between dry bulb and dew point temperatures. At saturation, wet bulb, dry bulb and dewpoint are equal. Wet bulb temperature alone is insufficient to determine state on a psychrometric chart, hence insufficient for determining specific humidity ratio (grains of water/lb. dry air) or any other property. Both wet and dry bulb temperatures are normally used to locate a point on a psychrometric chart. Wet bulb (adiabatic saturation) temperature is normally the lowest temperature obtained by evaporative cooling at ambient pressure.

Answer is (d) ●

THERMO 69

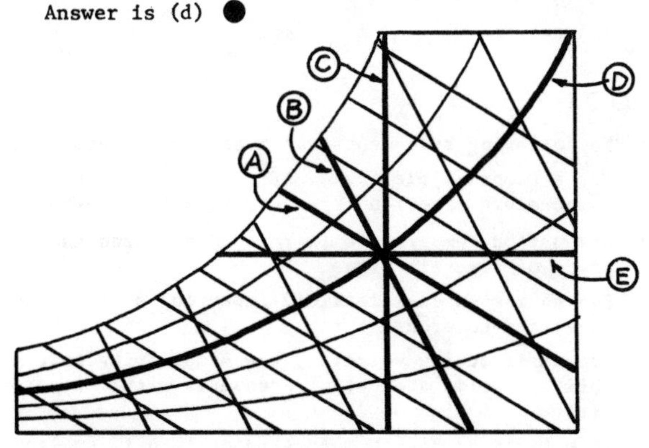

Line E of the usual psychrometric chart above represents a line of constant:

(a) relative humidity
(b) dew point or specific humidity
(c) enthalpy or wet bulb temperature
(d) specific volume
(e) dry bulb temperature

Sloped lines A are lines of constant wet bulb temperature and are almost lines of constant enthalpy.
Steeper lines B are lines of constant specific volume, cu ft wet air/lb dry air.
Vertical lines C are lines of constant dry bulb temperature.
Curved lines D are lines of constant relative humidity.
Horizontal lines E are lines of constant dew point, specific humidity and water vapor pressure.

Note: Psychrometric charts are only valid at one pressure, usually 14.7 psia, though they are usable with little error for barometric pressures between 29-31 " Hg. Answer is (b) ●

THERMO 70

A small plastic bag is filled with moist air at 50% relative humidity, sealed and placed in an environmental chamber whose temperature and pressure may be independently varied. The relative humidity in the bag will be lowered under which of the following conditions?

(a) pressure is increased
(b) pressure is decreased
(c) temperature is decreased
(d) temperature is decreased and pressure is increased
(e) none of the above conditions

Unless water condensation occurs at 100% R.H., the sealed bag contains a fixed mole fraction of water vapor in air.

$$\text{Relative Humidity} = \left\{ \frac{\text{mole fraction of water vapor in the mixture}}{\text{mole fraction of water vapor in a saturated mixture at the same temperature and pressure.}} \right\}$$

Vapor pressure of water is temperature dependent only, and increases with temperature. The denominator term of the above R.H. definition is increased with elevated temperature and decreased with elevated pressure. Thus: pressure increase raises R.H., pressure decrease lowers R.H., temperature decrease raises R.H., while simultaneous temperature decrease and pressure increase both raise R.H. Note that a temperature increase lowers R.H., but that choice was not given in the problem.

Answer is (b) ●

THERMO 71

Which of the following statements about stagnation properties is least correct?

(a) For a compressible fluid, KE change is converted to stagnation temperature rise, and pressure rise is isentropically calculated.

(b) Stagnation temperature is related to aerodynamic heating of leading edges of aircraft wings.

(c) For an incompressible fluid impacting a pitot tube, KE change may be converted to pressure rise.

(d) For gases at low velocity, the simplifying assumption of incompressibility yields stagnation pressures with only a small percentage error.

(e) In a fluid at Mach number greater than 1, an isentropic calculation of stagnation properties is valid.

Whenever a flow is brought to rest isentropically, a stagnation pressure arises. At Mach $>$ 1 a shock wave forms in front of the body and dissipates energy by a process which is not isentropic. Stagnation pressures produced at the body are lower than anticipated.

Answer (e) is false ●

THERMO 72

Which of the following statements about Mach number (M) is <u>false</u>?

(a) Mach number is the ratio of velocity to sonic velocity.

(b) Mach 1 is the maximum attainable velocity in a nozzle throat.

(c) Supersonic velocities ($M > 1$) are achievable in the diffuser section of a rocket nozzle if expansion ratio is great enough.

(d) Mach angle α, or angle the shock front makes with the velocity vector of a moving source, and Mach number are related by $M \sin \alpha = 1$.

(e) Mach no. has the dimensions of velocity.

Mach number, being a ratio of velocities, is dimensionless. All other statements are correct.

Answer (e) is false ●

THERMO 73

Sonic velocity (Mach 1) at 50,000 feet altitude in the standard upper atmosphere (-67.6°F and 1.68 psia) is nearest to:

(a) 880. ft/sec
(b) 971. ft/sec
(c) 995. ft/sec
(d) 1064. ft/sec
(e) 1117. ft/sec

Sonic velocity for ideal gases is temperature dependent only and may be calculated from

$C = \sqrt{k g_c R T}$ where C is sonic velocity in ft/sec.

$k = \frac{C_p}{C_v}$

$g_c = 32.17 \ lb_m \ ft/(lb_f \ sec^2)$

R = gas constant $ft \ lb_f/(lb_m \ {}^\circ R)$

T = temperature in °R.

$$C = \sqrt{1.40(32.17)(53.34)(392.4)} = 971 \text{ ft/sec}$$ ●

Answer is (b)

THERMO 74

Air flows through a 1 ft diameter circular duct at 20 psia and 80°F. A manometer connects a pitot tube, located on the centerline of the duct, with a static pressure tap upstream and reads a differential pressure of 3 inches of water. Determine mass flow rate of the air in lb/sec.

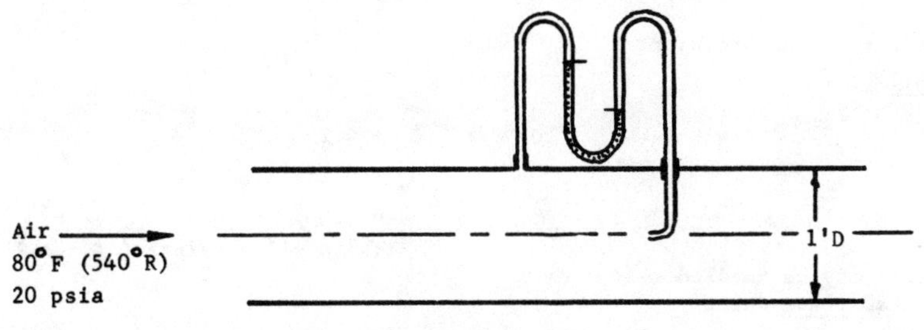

(1) Solution based on assumed incompressibility

Since differential pressure is trivial, this is acceptable.
(3" water = 0.25' water = 0.25(62.4) = 15.6 psf = 15.6/144 = 0.108 psi)

In the incompressible solution, KE is converted to an additional pressure energy indicated by the pitot tube, and velocity is determined.

Basis: 1 lb_m of fluid flowing.

$$\frac{V^2}{2g_c} = h = \frac{\Delta P}{\rho}$$

where h is additional head in feet = ft lb_f/lb_m
ρ is density of the fluid flowing in lb_m/ft^3
ΔP is differential pressure in lb_f/ft^2
V is velocity in ft/sec
g_c is 32.17 lb_m ft/(lb_f sec)

For air, $R = 53.34 \frac{ft\ lb_f}{lb_m\ °R}$ and density is the reciprocal of specific volume, $v = \frac{1}{\rho}$.

$$\rho = \frac{P}{RT} = \frac{20(144)}{53.34(540)} = 0.10\ lb/ft^3 = \text{density of air in the duct}$$

$$\frac{V^2}{2(32.17)} = \frac{15.6}{0.10} = 156. = h = \frac{\Delta P}{\rho}$$

This pressure energy (flow energy) is equivalent to a 156. foot (head) column of air at the T and P existing in the duct.

$$V = \sqrt{2(32.17)(156)} = 100\ ft/sec.$$

Duct cross-sectional area $= \frac{\pi}{4} d^2 = 0.785\ ft^2 = A$

Volume flow rate = 0.785 ft^2(100 ft/sec) = 78.5 cfs. = AV = q

Since $\dot{m} = q\rho$, $\dot{m} = 78.5 \frac{ft^3}{sec} \times 0.10 \frac{lb}{ft^3} = 7.85\ lb_m/sec.$ ●

(2) Solution based on compressible fluid

KE is converted to heat, resulting in a stagnation temperature, T_{stag} and this has a corresponding isentropic pressure rise, ΔP to stagnation pressure, P_{stag}.

Basis: 1 lb_m of fluid flowing

$$\frac{V^2}{2g_c J} = C_p \Delta T \qquad T_{stag} = T_1 + \Delta T = T_2$$
$$P_{stag} = P_1 + \Delta P = P_2$$

For an isentropic compression:

$$\left(\frac{P_2}{P_1}\right)^{\frac{k-1}{k}} = \frac{T_2}{T_1} \quad \text{where } k = \frac{C_p}{C_v}$$

For air: C_p = 0.240 BTU/lb°R
C_v = 0.171 BTU/lb°R
k = 1.40
(k is often referred to as γ)

These ratios may be handled using any consistent absolute units.

$$\left(\frac{20 + 0.108}{20}\right)^{\frac{1.40-1}{1.40}} = \frac{T_2}{540} \qquad (1.00540)^{0.2857} = \frac{T_2}{540}$$

$$1.00154 = \frac{T_2}{540} \qquad \therefore T_2 = 540.832°R \quad \text{or} \quad \Delta T = 0.832°R$$

To avoid inaccuracy, it is necessary to use as many digits as possible.

$V^2 = 2g_c J C_p \, \Delta T$ where J = 778 ft lb/BTU

$$V = \sqrt{2(32.17)(778)(0.240)(0.832)} = 100.\ \text{ft/sec.}$$

Note that this result agrees with the incompressible easier case where velocities are low and T_{stag} and P_{stag} are not much different from the duct temperature and pressure. The compressible solution is the only valid one at very high velocities.

Summary: V = 100 ft/sec $\therefore \dot{m}$ = 7.85 lb/sec ●

THERMO 75

During complete stoichiometric combustion of 1 lb. mole of methane (CH_4) with air, the number of lb. moles of nitrogen and other inerts that pass through the combustion zone is nearest to:

(a) 0.79
(b) 3.9
(c) 5.6
(d) 7.5
(e) 16.1

$CH_4 + 2O_2 \rightarrow CO_2 + 2H_2O$

1 lb. mole of methane requires 2 lb. moles of O_2 for complete combustion.

Air is 21% vol. O_2 and 79% vol. (N_2 + inerts).

2 lb. moles of O_2 is supplied by $\frac{2}{0.21}$ = 9.52 lb. moles of air, which contains 7.52 lb. moles of (N_2 + inerts). ●

Answer is (d)

THERMO 76

During combustion of hydrocarbon fuels with less than stoichiometric air, the products depend on the amount of air supplied and could be all of the following except:

(a) Unburned fuel, C, CO_2 and H_2O
(b) Unburned fuel, C, CO and H_2O
(c) C, CO, CO_2 and H_2O
(d) C, CO and H_2O
(e) CO, CO_2 and H_2O

Combustion of hydrocarbon fuels occurs stepwise in the following sequence:

first: all hydrogen to H_2O, leaving carbon

next: all carbon to CO

last: all CO to CO_2

This is summarized in the table below:

Amount of Air	Products from Combustion Zone				
	Unburned Fuel	Carbon	CO	CO_2	H_2O
excess air				X	X
stoichiometric				X	X
slightly fuel rich			X		X
less than stoichiometric air ↓		X	X		X
	X	X	X		X
very fuel rich	X	X			X

As a practical matter it is difficult to obtain a CO-free stack gas due to slow kinetics of CO combustion to CO_2, and due to equilibrium in the water gas shift reaction.

Choice (a) in the problem does <u>not</u> occur ●

THERMO 77

(a) Write the combustion equation for a heavy oil whose composition averages $C_{17}H_{36}$ with complete combustion in theoretical air.

(b) Determine the pounds of theoretical air per pound of $C_{17}H_{36}$ fuel.

(a) During stoichiometric combustion all hydrogen is burned to water and carbon to CO_2.

$$C_{17}H_{36} + 26O_2 \longrightarrow 17CO_2 + 18H_2O \quad ●$$

From the calculation below, 97.8 (N_2 + inerts) pass through unreacted.

(b) Basis of calculation: 1 lb. mole of $C_{17}H_{36}$

26 lb. moles of O_2 are required. Air is 21% vol. (21 mole%) O_2 and 79% vol. (79 mole%) N_2 + inerts.

$$\frac{26 \text{ lb. moles } O_2}{21} = \frac{X \text{ lb. moles } (N_2 + \text{inerts})}{79} = \frac{Y \text{ lb. moles air}}{100}$$

X = 97.8 lb. moles (N_2 + inerts) that pass through the combustion zone

and Y = 123.8 lb. moles air required.

Since mol. wt. of $C_{17}H_{36}$ = 17(12) + 1(36) = 240

and the mol. wt. of air is 28.95

$$\frac{\text{lb air}}{\text{lb fuel}} = \frac{123.8(28.95)}{1(240)} = 14.9 \text{ air/fuel wt. ratio} \quad ●$$

THERMO 78

Gasoline has a heat of combustion of 21,000 BTU/lb and a specific weight of 6.17 lb/gal. Food has an average heat of combustion of roughly 6000 BTU/lb, and the average human's rate of energy comsumption is approximately 12,000 BTU/day. The automobile's average rate of fuel consumption in city driving is 3 gallons per hour.

Given that the equation for the combustion of both gasoline and food is

Fuel + Oxygen = Carbon Dioxide + Water,

and given that the chemical formulas for gasoline and food are C_8H_{18} and $C_6H_{12}O_6$, respectively, compute the ratio of the amount of air used per hour by the automobile to the amount used by an average human.

Basis of calculation: 1 hour

(1) Automobile

Stoichiometric equation: $C_8H_{18} + 12.5\ O_2 \rightarrow 8CO_2 + 9H_2O$

1 lb. mole of fuel, mol. wt. = 8(12) + 18(1) = 114 lb. fuel, requires 12.5 lbmoles of oxygen

Fuel consumption = 3 gal/hr x 6.17 lb/gal = 18.51 lb/hr.

Oxygen requirement is obtained by ratio:

$$\frac{18.51}{114} = \frac{\text{lbmoles oxygen}}{12.5}$$

Oxygen = 2.030 lb moles/hr.

(2) Human

Stoichiometric equation: $C_6H_{12}O_6 + 6O_2 \rightarrow 6CO_2 + 6H_2O$

1 lb. mole of food, mol. wt. = 6(12) + 12(1) + 6(16) = 180 lb. food, requires 6 lb. moles of oxygen.

$$\text{Food consumption} = \frac{12{,}000\ \text{BTU/day}}{6{,}000\ \text{BTU/lb.}} \times \frac{1\ \text{day}}{24\ \text{hr}} = 0.0833\ \text{lb/hr.}$$

Oxygen requirement is obtained by ratio:

$$\frac{0.0833}{180} = \frac{\text{lbmoles oxygen}}{6}$$

Oxygen = 0.00278 lb moles/hr.

(3) Comparison

$$\frac{\text{Auto Air}}{\text{Human Air}} = \frac{\text{Auto Oxygen}}{\text{Human Oxygen}} = \frac{2.030}{0.00278} = 731.$$

THERMO 79

Propane (C_3H_8) is burned in a heater with 20% excess air at ambient conditions (68°F, 14.7 psia, 30% relative humidity). Assume complete combustion to water and carbon dioxide.

Per pound mole of propane, calculate:

(a) volume of stack gases cooled to 800°F.

(b) composition of stack gases (vol.%).

(c) dew point of stack gases.

Calculation basis: 1 lb. mole of C_3H_8

(1) Stoichiometric equation: $C_3H_8 + 5O_2 \longrightarrow 3CO_2 + 4H_2O$

5 lb. moles O_2 are required for complete combustion. At 20% excess air, 6 lb. moles enter combustion zone accompanied by some moisture and much nitrogen plus inerts.

(2) Calculation of moisture and N_2 plus inerts carried by intake air

From the saturated steam table, water vapor pressure at 68°F is 0.339 psia; this corresponds to 100% relative humidity. At 30% R.H. the water vapor pressure is (0.30)(0.339) = 0.102 psia.

Mole fraction of water vapor is $\frac{0.102}{14.7} = 0.00695$, or 0.695% and mole fraction of dry air is 0.993.

Alternately, from a psychrometric chart, the water contained in intake air is 31. grains/(lb. dry air). At 7000 grains/lb., this is 0.0044 lb. H_2O/(lb. dry air).

$$\text{Mole fraction of water vapor} = \frac{\text{moles water vapor}}{\text{moles dry air + moles water vapor}}$$

$$= \frac{\frac{0.0044}{18}}{\frac{1}{28.9} + \frac{0.0044}{18}} = 0.070$$

Mole fraction of dry air = 1 - 0.070 = 0.993

6 lb. moles O_2 are carried by $\frac{100}{21}(6)$ = 28.6 lb. moles dry air or by $\frac{28.6}{0.993}$ = 28.8 lb. moles moist air containing 0.2 lb. moles H_2O.

Nitrogen + inerts carried = 28.6 - 6.0 = 22.6 lb. moles.

(3) Summary: lb. moles entering and leaving combustion zone

	C_3H_8	O_2	N_2 plus inerts	H_2O	CO_2	Total
Entering	1.0	6.0	22.6	0.2	0	29.8
Leaving	0	1.0	22.6	0.2+4.0 = 4.2	3.0	30.8

Note from the stoichiometric equation that 6 moles of reactants form 7 moles of products - justifying change in moles in and out above.

(4) Computation of stack gas volume

Use ideal gas law PV = nRT, where $R = \frac{1545 \text{ ft lb}_f}{\text{lb}_m \text{ mole °R}}$, to calculate volume of 30.8 moles stack gases at 800 F (1260°R) and 14.7 psia.

$14.7(144)V = 30.8(1545(1260) \quad \therefore V = 28{,}300 \text{ ft}^3$

Alternate calculation based on 1 lb.mole = 359 ft^3 at 32°F, 14.7 psia:

$$30.8(359)\frac{1260°R}{492°R} = 28{,}400 \text{ ft}^3$$

(5) Stack gas composition (vol.% = mole %)

Component	Excess O_2	N_2 plus inerts	CO_2	H_2O	Total
lb. moles	1.0	22.6	3.0	4.2	30.8
vol. %	3.25	73.35	9.75	13.65	100.00 ●

(6) Dew point of stack gas

Determine temperature where vapor pressure of water equals partial pressure of water in stack gas.

From above table, vol. % H_2O = 13.65 mole fraction = 0.1365

Water partial pressure in stack gas = (14.7 psia)(0.1365) = 2.01 psia.

Consult saturated steam table for vapor pressure of water at various temperatures. Find:

120°F 1.692 psia
130°F 2.221 psia

Although vapor pressure does not vary linearly with temperature, linear interpolation is adequate over a short temperature range and is as accurate as the problem input data justify. Conclude that dew point = 126°F. ●

Summary of results

(a) Stack gas volume = 28,300 cf/lb.mole C_3H_8 ●
(b) Stack gas composition, vol. % (see table above) ●
(c) Stack gas dew point = 126°F ●

THERMO 80

Theoretically, how many pounds of air are required for complete combustion of 10 pounds of ethane gas (C_2H_6), and how many pounds of water vapor are produced in the products of combustion? Assume the air contains 23% oxygen by weight.

Write and balance the combustion equation

$$2C_2H_6 + 7O_2 \longrightarrow 4CO_2 + 6H_2O$$

The molecular weights must be calculated

C = 12 H = 1 O = 16

Therefore C_2H_6 = 30 O_2 = 32 CO_2 = 44 H_2O = 18

Thus 2 moles of C_2H_6 (2 x 30 lbs) plus 7 moles of O_2 (7 x 32 lbs) yield 4 moles of CO_2 (4 x 44 lbs) plus 6 moles of H_2O (6 x 18 lbs).

Therefore 10 lbs of C_2H_6 require

$$\frac{10}{60}(7 \times 32) = 37.33 \text{ lbs of oxygen}$$

Since air is 23% oxygen by weight,

$$\text{pounds of air required} = \frac{37.33}{0.23} = 162 \text{ pounds}$$ ●

10 lbs of C_2H_6 will produce

$$\frac{10}{60}(6 \times 18) = 18 \text{ lbs of water vapor}$$ ●

THERMO 81

If the heat of combustion of kerosene ($C_{10}H_{22}$) is 1625 K cal/mole and kerosene costs 25 cents per gallon, what is the cost of 25 K cal obtainable by burning kerosene? Assume complete combustion of kerosene, the density of kerosene is 0.80 grams per c.c., and that one fluid ounce is equal to 30 c.c.

Atomic Weights   O - 16   C - 12   H - 1

mole = gram-molecular weight

number of moles burned $= \frac{25}{1625} = \frac{1}{65}$ mole kerosene

$C_{10}H_{22}$ has a molecular weight of 10(12) + 22(1) = 142

kerosene burned $= \frac{1}{65} \times 142 = 2.185$ grams

$$\text{kerosene cost} = \frac{2.185 \text{ gms}}{0.8 \frac{\text{gms}}{\text{cc}}} \times \frac{1}{30 \frac{\text{cc}}{\text{fl oz}}} \times \frac{1}{32 \frac{\text{fl oz}}{\text{quart}}} \times \frac{1}{4 \frac{\text{qt}}{\text{gal}}} \times 25 \frac{\text{cents}}{\text{gal}}$$

$= 0.0178$ cents ●

THERMO 82

Given the following data for propane:

Formula	C_3H_8
Weight (liquid)	4.24 lb/gal
Specific Volume	8.55 ft^3 gas/lb

REQUIRED: Compute the following:

(a) lbs O_2 per lb of gas for complete combustion

(b) lbs CO_2 produced per lb of gas burned

(c) lbs H_2O produced per lb of gas burned

First, write the balanced chemical equation.

$$C_3H_8 + 5O_2 \longrightarrow 3CO_2 + 4H_2O$$

Knowing the atomic weights, compute the molecular weights.

44   32   44   18   (C = 12, O = 16, H = 1)

So, 44 pounds of C_3H_8 combine with 5 x 32 = 160 pounds of oxygen to form 3 x 44 = 132 pounds of CO_2 plus 4 x 18 = 72 pounds of H_2O.

(a) $\frac{160}{44} = 3.64$ pounds of O_2 per pound of C_3H_8 ●

(b) $\frac{132}{44} = 3.0$ pounds of CO_2 per pound of C_3H_8 ●

(c) $\frac{72}{44} = 1.64$ pounds of H_2O per pound of C_3H_8 ●

THERMO 83

The temperature difference between the two sides of a solid rectangular slab of area A and thickness L as shown below is ΔT. The heat transferred through the slab by conduction in time t is proportional to:

(a) $AL\Delta T\,t$

(b) $AL\,\dfrac{\Delta T}{t}$

(c) $AL\,\dfrac{t}{\Delta T}$

(d) $\dfrac{A\Delta T\,t}{L}$

(e) $\dfrac{A(\Delta T)^2 t}{L}$

Heat transferred through slab by conduction in time t is:

$$Q = kA\Delta T \quad \text{where} \quad k = \frac{\text{BTU ft}}{\text{hr ft}^2\ {}^\circ\text{F}}$$

$$Q = k\left(\frac{\text{BTU ft}}{\text{hr ft}^2\ {}^\circ\text{F}}\right) A\,\frac{(\text{ft}^2)(\Delta T)^\circ\text{F t hrs}}{\text{L ft}} = \text{BTU} = \frac{k(A\Delta T\,t)}{L}$$

k is the coefficient of thermal conductivity of the material, hence heat transfer in a given material is proportional to the other variables.

Answer is (d) ●

THERMO 84

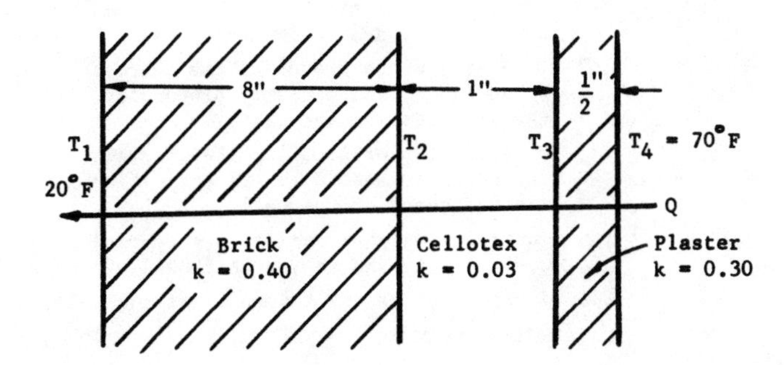

Given the composite wall shown with steady state outer temperature $T_1 = 20^\circ$F and inner temperature $T_4 = 70^\circ$F, which of the following is nearest to T_3 in °F?

(a) 27.
(b) 38.
(c) 46.
(d) 58.
(e) 69.

At steady state the same Q flows across each material and temperatures descend in direct relation to thermal resistance (reciprocal of conductivity).

Resistance of brick $= \frac{x}{k} = \frac{0.667 \text{ ft}}{0.40 \text{ BTU/(ft}^2\text{°F/ft)}} = 1.67 \text{ ft}^2\text{°F/(BTU)}$

Resistance of Cellotex $= \frac{x}{k} = \frac{0.083 \text{ ft}}{0.03} = 2.77$

Resistance of plaster $= \frac{x}{k} = \frac{0.042 \text{ ft}}{0.30} = 0.14$

Total Resistance $= 4.58 \text{ ft}^2\text{°F/(BTU)}$

$$Q = \frac{\Delta T_{overall}}{\text{total resistance}} = \frac{\Delta T \text{ for plaster}}{\frac{x}{k} \text{ for plaster}} = \ldots$$

$$Q = \frac{50}{4.58} = \frac{T_4 - T_3}{0.14} = \frac{T_3 - T_2}{2.77} = \frac{T_2 - T_1}{1.67} = 10.85 \text{ BTU/hr.}$$

$T_4 - T_3 = 1.5\text{°F}$ Since $T_4 = 70.\text{ °F}$, $T_3 = 68.5\text{°F}$ ●

$T_3 - T_2 = 30.3\text{°F}$ Since $T_3 = 68.5\text{°F}$, $T_2 = 38.2\text{°F}$

$T_2 - T_1 = 18.2\text{°F}$ Since $T_2 = 38.2\text{°F}$, $T_1 = 20\text{°F}$ (in agreement with given data)

Answer is (e)

THERMO 85

The rate of heat transfer through a given section of a uniform wall for a given temperature difference is:

(a) Directly proportional to the thermal conductivity and to the thickness of the wall.

(b) Inversely proportional to the thermal conductivity and directly proportional to the thickness of the wall.

(c) Directly proportional to the thermal conductivity and inversely proportional to the thickness of the wall.

(d) Inversely proportional to the thermal conductivity and to the thickness of the wall.

(e) Independent of the thickness of the wall.

$$Q = \frac{k\,A(\Delta T)}{L}$$

where k = conductivity
A = surface area
ΔT = temperature difference
L = wall thickness

Answer is (c) ●

THERMO 86

A laboratory furnace has a working chamber having the dimensions: 6 inches by 10 inches by 15 inches. The walls are 4 inches thick on all sides and are made of a refractory brick with a thermal conductivity of 0.25 $BTU\text{-}ft/hr\text{-}ft^2\,°F$.

Assume that the furnace chamber is held at 1900°F and the outside surface is estimated to be at 300°F. Determine an estimate of the electrical power consumption of the furnace.

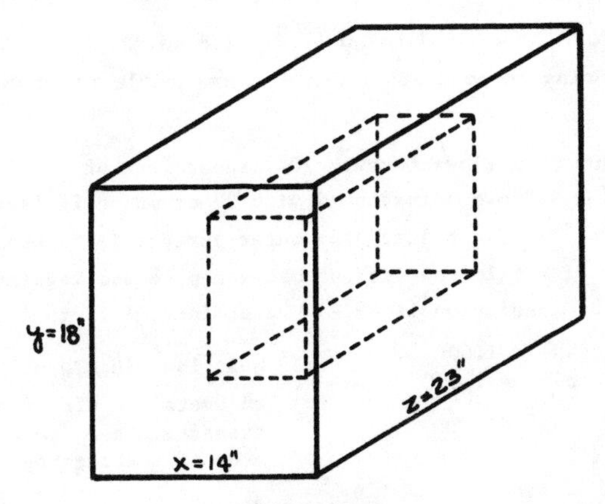

Inner surface area = 2 sides + 2 tops + 2 fronts
= 2(6 x 15) + 2(10 x 15) + 2(6 x 10) = 600 in^2

Outer surface area = 2(14 x 23) + 2(18 x 23) + 2(14 x 18) = 1976 in^2

For a problem where heat flow is always perpendicular to the surface, a log-mean surface area yields the same result as an integration.

When applied to this problem the log mean surface area will yield high results for conduction, because heat flow at edges and corners is not perpendicular to the surface. This is intuitively seen or would be shown by a network analysis beyond the scope of this review.

$$\text{log mean surface area} = \frac{A_{outer} - A_{inner}}{\ln \frac{A_{outer}}{A_{inner}}} = \frac{1976\ in^2 - 600\ in^2}{\ln \frac{1976}{600}} = 1152\ in^2$$

$$= \frac{1152}{144} = 8.0\ ft^2$$

$$Q = \frac{kA\Delta T}{L} = \frac{0.25(8.0)(1900 - 300)}{0.333} = 9600\ \text{BTU/hr} \quad \text{(high result)}$$

Alternately, an empirical formula of Langmuir for hollow rectangular parallelpipeds with uniform wall thickness has been used:

$$q = \left[\frac{A}{L} + 0.54 \sum \text{lengths inside edges} + 1.2\ L\right] k_{avg}(T_1 - T_2)$$

where A = internal surface area = 600 in^2 = 4.17 ft^2

L = wall thickness = 4 inches = 0.333 ft.

$\sum e$ = summation of internal edges = 124 inches = 10.33 ft.

This equation is claimed to be applicable where each inside edge is between 0.2 - 2.0 times wall thickness (this problem has inner edges 1.5 to 3.75 times wall thickness and the equation can only yield low results since Langmuir's model has a larger fraction of outer edge and corner areas than this furnace).

$$q = \left[\frac{4.17}{0.33} + 0.54(10.33) + 1.2(0.333)\right] 0.25(1900 - 300) = 7396.\ \text{BTU/hr.}$$

(equivalent to 2.17 Kw)

We believe this number to be low.

Power Requirement: Since 1 Kw = 3413 BTU/hr., the steady state power input is:

$$\text{above } \frac{7400}{3413} = 2.17 \text{ Kw}, \qquad \text{and below } \frac{9600}{3413} = 2.8 \text{ Kw.}$$ ●

Alternate solutions using shape factors yield a comparable range of answers.

THERMO 87

Saturated steam at 100 psig flows through 200 linear feet of 1.5" schedule 40 pipe (1.60" i.d. and 1.90" o.d.) insulated with 3" of magnesia lagging ($k = 0.042$ BTU/(hr ft^2°F/ft). Assume pipe outer surface is at steam temperature and ignore existence of a film coefficient between pipe and lagging. Outer surface temperature of insulation is 80°F. Determine:

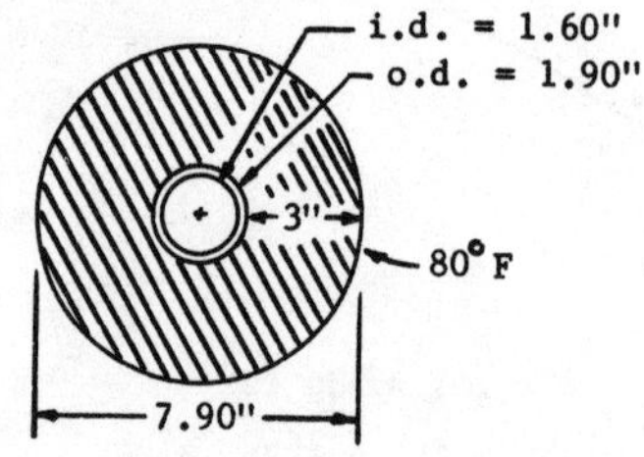

(a) heat loss in BTU/hr.

(b) an overall coefficient for heat transfer, based on outer surface area of the lagging.

Steam: 100 psig = 114.7 psia

From saturated steam tables, temperature = 338.°F

(341.3°F at 120 psia, and 334.8°F at 110 psia, interpolated to 337.8°F)

This radial flow of heat can be calculated from a handbook formula derived from the basic conduction equation:

$$q = -kA\frac{dT}{dx}$$

Rearrange, use annular area elements from inner to outer radius and integrate from inner $r_1 = 0.95"$ to outer $r_2 = 3.95"$ radius. Area of an annular element $A = 2\pi r$, and $dx = dr$.

$$q\int_{r_1}^{r_2}\frac{dx}{A} = -k\int_{T_1}^{T_2} dT \qquad q = \frac{-k_{avg}(T_2 - T_1)}{\int_{r_1}^{r_2}\frac{dr}{2\pi r}} = \frac{2\pi k_{avg}(T_1 - T_2)}{\ln\frac{r_2}{r_1}}$$

$$q = \frac{2\pi k_{avg}(T_1 - T_2)}{2.303 \log\frac{r_2}{r_1}} = \frac{2\pi(0.042)(338 - 80)}{2.303 \log\frac{3.95}{0.95}} = \frac{68.0}{1.42} = 47.8 \text{ BTU/hr per linear foot}$$

$$= 9560 \text{ BTU/hr for 200' length}$$ ●

The outer surface area of 200 linear feet:

$$A = 2\pi rL = 2\pi\left[\frac{3.95}{12}\right](200) = 413 \text{ ft}^2$$

An overall coefficient based on this area is found from

$q = UA\Delta T,$ $9560 = U(413.)(338 - 80)$

$$U = 0.090 \text{ BTU/(hr ft}^{2}{}^{\circ}\text{F)}$$ ●

THERMO 88

The heat loss per hour through 1 sq ft of furnace wall 18" thick is 520 Btu's. The inside wall temperature is 1900°F, and its average thermal conductivity is 0.61 Btu/hr ft °F.

REQUIRED: Determine the outside surface temperature of the wall.

$q = k \frac{A}{L}(t_1 - t_2)$ where $t_1 = 1900°F$

$k = 0.61$ Btu/hr ft °F

$\frac{q}{A} = 520$ Btu/hr

$L = 1.5$ ft

Solving for t_2

$$t_2 = -\frac{q}{A}\frac{L}{k} + t_1 = -520\frac{1.5}{0.61} + 1900 = -1279 + 1900 = 621°F$$

Outside surface temperature of the wall = 621°F. ●

THERMO 89

Heat is transferred by conduction from left to right through the composite wall shown in the drawing below:

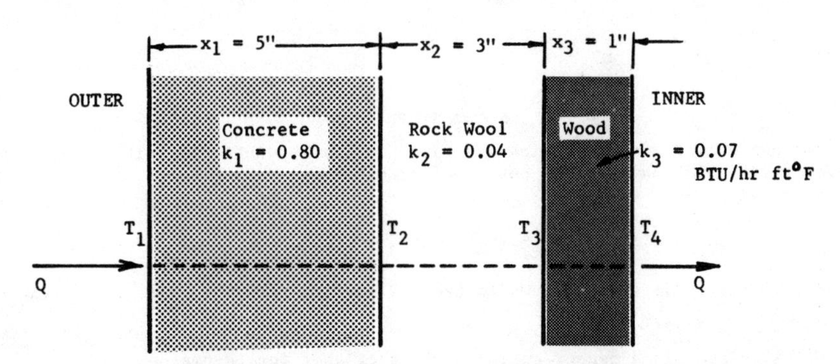

Assume the three materials are in good thermal contact and no significant film coefficients exist at any of the interfaces. Determine which of the following numbers most nearly represents the overall coefficient U in BTU/hr ft^2°F.

(a) 0.04
(b) 0.13
(c) 0.35
(d) 0.91
(e) 1.92

Overall coefficient U, thermal conductivity, $\frac{k}{x}$, and film coefficient, h, are reciprocals of their thermal resistances. Thermal resistances in series are handled analogously to series electrical resistances, hence

$$U \text{ in BTU/(hr ft}^2{}^\circ\text{F)} = \frac{1}{\frac{x_1}{k_1} + \frac{x_2}{k_2} + \frac{x_3}{k_3} + \cdots}$$

Overall coefficient U is used in a simplified conduction equation $Q = UA\Delta T$.
In this problem:

$$U = \frac{1}{\frac{5/12}{0.80} + \frac{3/12}{0.04} + \frac{1/12}{0.07}} = \frac{1}{0.52 + 6.25 + 1.19} = 0.126 \text{ BTU/(hr ft}^2{}^\circ\text{F)} \bullet$$

Answer is (b)

THERMO 90

Given an inner wall at 80°F and an outer wall exposed to ambient wind and surroundings at 40°F. Film coefficient, h, for convective heat transfer at a 15 mph wind is about 7 BTU/(hr ft²°F).

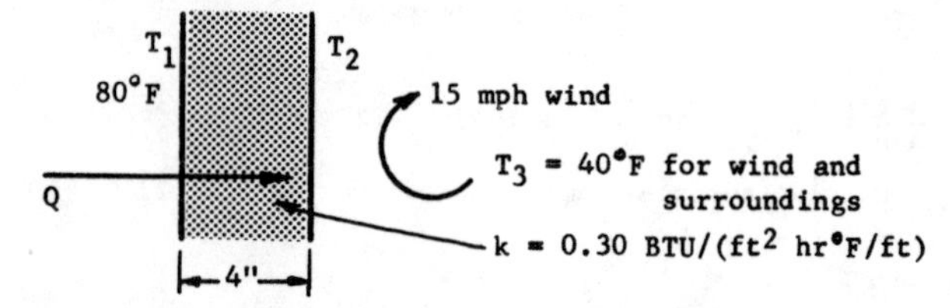

Ignoring any radiation losses calculate an overall coefficienct for the conduction and convection losses.

(a) 0.14
(b) 0.80
(c) 1.25
(d) 7.1
(e) 8.2

Since conduction and convection are based on ΔT, absolute temperatures are not required. For steady state, heat conducted through wall = heat lost by convection:

$$Q = \frac{kA(T_1 - T_2)}{x} = hA(T_2 - T_3) \quad (1)$$

Q can similarly be expressed by an overall coefficient

$$Q = UA(T_1 - T_3) \quad (2)$$

U is calculated in a manner analogous to that used for thermal conductivities in series.

$$U = \frac{1}{\frac{1}{h_1} + \frac{x_1}{k_1} + \cdots} \quad (3)$$

$$U = \frac{1}{\frac{1}{7} + \frac{4/12}{0.30}} = \frac{1}{0.143 + 1.11} = 0.80 \text{ BTU/(hr ft}^2{}^\circ\text{F)} \bullet$$

Alternate Solution

If the question had asked for T_2, equation (1) would have given it. Similarly the solution for Q from (1) would yield U from (2):

$$Q = \frac{0.30}{\frac{4}{12}}(80 - T_2) = 7(T_2 - 40)$$

$Q = 72 - 0.9T_2 = 7T_2 - 280 \qquad \therefore T_2 = \frac{352}{7.9} = 44.5°F$ and $Q = 32$ BTU/(hr ft^2)

$Q = UA(T_1 - T_3) = 32 = U(1)(80 - 40) \qquad U = \frac{32}{40} = 0.80$ BTU/(hr ft^2°F) ●

Note that this alternate approach becomes inoperative when more than one film coefficient and one conductivity are involved, but lacking information on the interface temperatures.

Answer is (b)

THERMO 91

A metal object at 120°F is set on an insulating pad to cool. The temperature initially falls from 120°F to 100°F in 12 minutes. Surroundings are at 65°F. Find time required for that object to continue to cool from 98°F to 80°F. Assume negligible conduction and radiation losses in both cases.

rate of heat loss from block = rate of convective heat transfer

$-\rho CV \frac{d\theta}{dt} = hA\theta$, where ρ = density, C = heat capacity, V = volume, t = time,
θ = temperature difference with surroundings,
h = convective heat transfer coefficient, and A = area.

Rearrange and integrate using the first cooling limits:

$$\ln \theta \Big]_{\theta_1 = 55°F}^{\theta_2 = 35°F} = \left(-\frac{hA}{\rho CV}\right) t \Big]_{t_1 = 0min.}^{t_2 = 12min.} \qquad \ln\left(\frac{35}{55}\right) = \left(-\frac{hA}{\rho CV}\right)(12 - 0)$$

Solve for the unspecified constants: $\left(-\frac{hA}{\rho CV}\right) = -.0377 \text{ min}^{-1}$

Insert the new problem data and re-evaluate the integral. $\theta_3 = 98-65$, $\theta_4 = 80-65$, $t_4 - t_3$ = unknown.

$$\ln\left(\frac{15}{33}\right) = -.0377\,(t_4 - t_3) \qquad t_4 - t_3 = 20.9 \text{ min.}$$ ●

This demonstrates the exponential nature of temperature change with time, where the heat transfer rate by conduction or by convection varies solely with temperature difference, i.e. in the absence of heat generated in, or lost from the object in other ways.

THERMO 92

A shell and tube brine cooler cools 150 gallons of brine per minute from 16°F to 12°F, using ammonia at 5°F. The effective outside area of the tubes is 310 ft^2. The brine has a specific gravity of 1.2 and a specific heat of 0.70.

(a) Compute the rating of the cooler in tons of refrigeration.

(b) Compute the coefficient of heat transfer, U, in BTU/(hr ft^2°F).

(a) Heat of Cooling = $Q_c = \dot{m}_B \, C_{p_B}(T_1 - T_2)$

where $\dot{m}_B$ = brine mass flow rate

C_{p_B} = brine specific heat = 0.70

T_1 = brine inlet temperature = 16°F

T_2 = brine outlet temperature = 12°F

$$\frac{150 \text{ gal}}{\text{min}} \times \frac{\text{cu ft}}{7.48 \text{ gal}} = 20 \text{ ft}^3/\text{min}$$

1 ton of refrigeration = 200 Btu/min = 12,000 Btu/hr

$$\dot{m}_B = \frac{62.4 \text{ lb}}{\text{cu ft}} \times 1.2 \times \frac{20 \text{ cu ft}}{\text{min}} \times \frac{60 \text{ min}}{\text{hr}} = 90{,}000 \text{ lbm/hr}$$

$$Q_c = 90{,}000 \times 0.70(16° - 12°) = 252{,}000 \text{ Btu/min}$$

$$\text{Tons of Refrigeration} = \frac{252{,}000}{12{,}000} = 21. \bullet$$

(b) $U = \dfrac{q}{A_T \, \Delta t_m}$ where q = heat transfer in Btu/hr

A_T = effective area of the tubes in sq ft

Δt_m = log mean temp difference

$$\Delta t_m = \frac{(\Delta t_0)_a - (\Delta t_0)_b}{\text{Ln}\left[(\Delta t_o)_a / (\Delta t_0)_b\right]}$$

$(\Delta t_0)_a = 16° - 5° = 11°\text{F}$

$(\Delta t_0)_b = 12° - 5° = 7°\text{F}$

$$= \frac{11° - 7°}{\text{Ln}(11/7)} = \frac{4°}{0.45} = 8.9°\text{F}$$

We are using the logarithmic mean temperature difference which applies to countercurrent heat exchange; ideally a correction should be applied as this is a shell and tube exchanger.

$$U = \frac{q}{A_T \Delta t_m} = \frac{252{,}000 \text{ BTU/hr}}{310 \text{ ft}^2 (8.9°\text{F})} = 91.4 \text{ BTU/(hr ft}^2 °\text{F)} \bullet$$

THERMO 93

Which of the following is not a usual expression of the power/unit area Stefan-Boltzmann constant for black body radiation?

(a) 1.36×10^{-12} cal/(sec cm^2°K^4)

(b) 5.68×10^{-5} ergs/(sec cm^2°K^4)

(c) 5.68×10^{-8} watts/(m^2°K^4)

(d) 0.171×10^{-8} BTU/(ft^2 hr°R^4)

(e) 5.68×10^{-8} coulombs/(sec m^2°K^4)

All are numerically correct conversion of the constant in terms of power per unit area.

The electrical units of (e) are not recognized as appropriate, although they are numerically equivalent to a correct constant of

5.68×10^{-8} coulombs/(sec m^2°K^4)

(e) is not a usual expression ●

8

Electrical Circuits

ELECTRICAL 1

In the transistor circuit shown below, the emitter-to-base voltage drop is 0.50 volt. What is the base current?

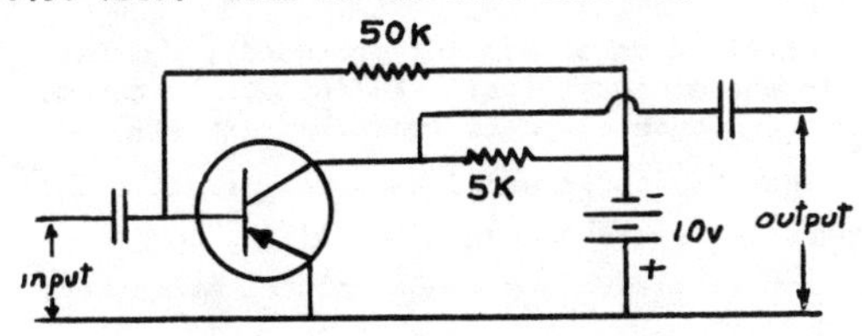

(a) 0.17 milliampere
(b) 0.19 milliampere
(c) 0.20 milliampere
(d) 2.09 milliampere
(e) 0.17 ampere

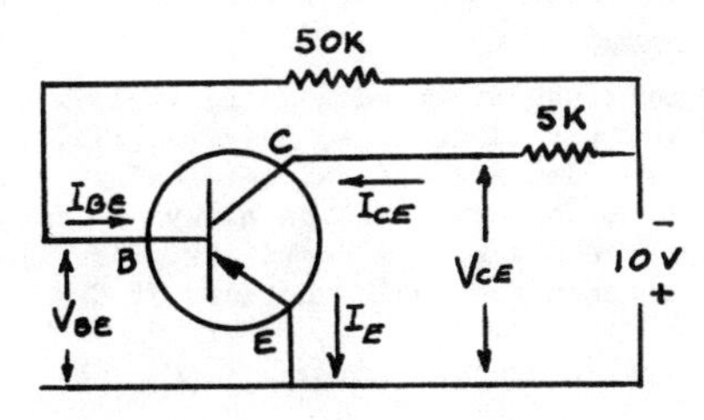

This pnp common emitter amplifier circuit has the current flows shown:

$I_{emitter} = I_{BE} + I_{CE}$

Since $V_{BE} = 0.5$ volt, voltage across the 50K resistor is 9.5 volts.

Apply Ohm's law: $I_{BE} = \frac{9.5}{50,000} = 19 \times 10^{-5} = 0.19$ ma ●

Answer is (b)

ELECTRICAL 2

A single-phase inductive load takes 50 kva at 0.60 power factor lagging. The amount of capacitance required to improve the power factor to 1.0 is:

(a) 10 kilovars
(b) 20 kilovars
(c) 22.5 kilovars
(d) 30 kilovars
(e) 40 kilovars

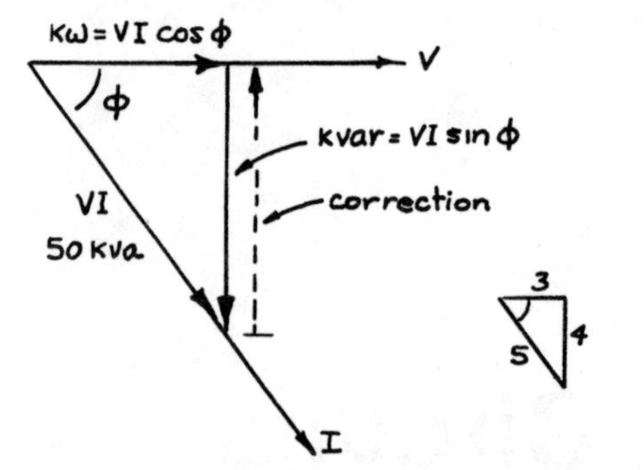

pf = 0.6 = cos ϕ

$\phi = 53.1^{\circ}$

sin ϕ = 0.8

VI sin ϕ = 40 kvar

VI cos ϕ = 50(0.6) = 30 kw

40 kvar of capacitors placed in parallel with the inductive load will improve the power factor to 1.0 (reduce ϕ to 0° lag), give a net reactive power of 0 kvar, reduce VI from 50 to 30 kva, while retaining the useful in-phase power at 30 kw. ●

Answer is (e)

ELECTRICAL 3

NPN or PNP transistors may be utilized in common (or grounded): emitter, base or collector circuits. In these circuits all the following statements relative to their comparable vacuum tube circuits are true, EXCEPT:

(a) Common emitter is comparable to grounded cathode circuit.
(b) Common base is comparable to grounded grid amplifier.
(c) Common collector is comparable to a grounded plate, or cathode follower circuit.
(d) Common base is comparable to a rectifier diode.
(e) Common emitter is one of the more widely used amplifier circuits.

Common (or grounded) element configurations are used comparatively with triode vacuum tube circuits: where emitter is cathode, base is grid, and collector is plate. The common or grounded element is contained in both the input and output circuits. Emitter to base junction always is forward biased, and base to collector is reverse biased. A rectifier diode contains a single P-N junction rather than the dual junctions of NPN or PNP transistors.

The false statement is (d) ●

ELECTRICAL 4

Solve for R in the circuit given below.

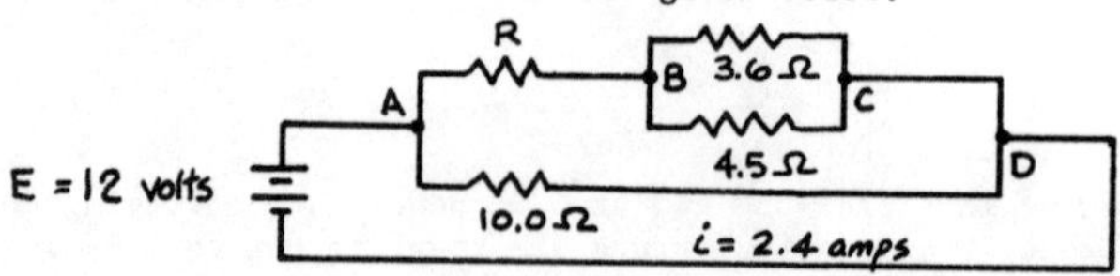

For resistances in parallel $\frac{1}{R_{Total}} = \frac{1}{R_1} + \frac{1}{R_2} + \frac{1}{R_3} + \cdots$

For two resistances, this reduces to

$$R_{1-2} = \frac{R_1 R_2}{R_1 + R_2}$$

$$R_{BC} = \frac{(3.6)(4.5)}{3.6 + 4.5} = \frac{16.2}{8.1} = 2 \text{ ohms}$$

$$R_{equiv} = R_{AD} = \frac{(R + R_{BC})(10)}{R + R_{BC} + 10} = \frac{10(R + 2)}{R + 12}$$

$$R_{equiv} = \frac{E}{I} = \frac{12}{2.4} = 5 \qquad 5 = \frac{10(R + 2)}{R + 12} \qquad 10R + 20 = 5R + 60$$

$$5R = 40$$

$$R = 8 \text{ ohms} \; ●$$

ELECTRICAL 5

Given values of the Y-connected impedances Z_1, Z_2 and Z_3 shown, find Z_A, Z_B, and Z_C of the equivalent Δ-connected circuit.

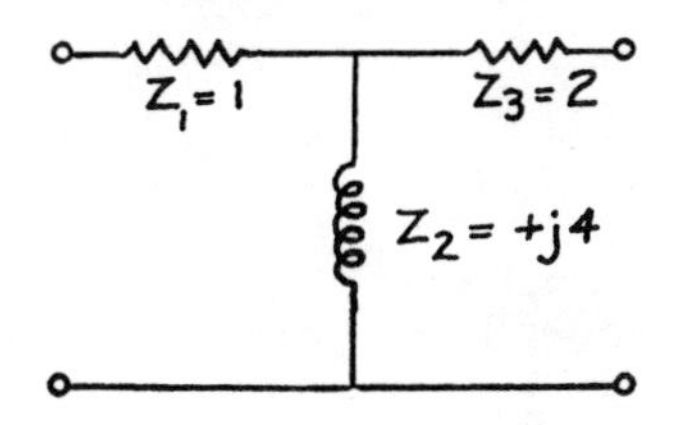

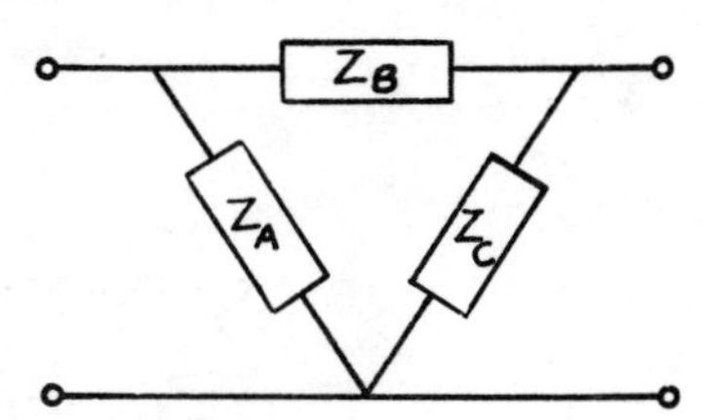

Let $S = Z_1Z_2 + Z_1Z_3 + Z_2Z_3 = 1(+j4) + 1(2) + (+j4)(2) = 2 + j12$

Then:

$$Z_A = \frac{S}{Z_3} = \frac{2 + j12}{2} = 1 + j6 \; ●$$

$$Z_B = \frac{S}{Z_2} = \frac{2 + j12}{+j4} = \frac{(2 + j12)(-j4)}{(+j4)(-j4)} = \frac{+48 - j8}{+16} = 3 - j0.5 \; ●$$

$$Z_C = \frac{S}{Z_1} = \frac{2 + j12}{1} = 2 + j12 \; ●$$

To check, convert the Δ-circuit back to the Y-circuit.

Let $T = Z_A + Z_B + Z_C = (1 + j6) + (3 - j0.5) + (2 + j12) = 6 + j17.5$

Then:

$$Z_1 = \frac{Z_AZ_B}{T} = \frac{(1 + j6)(3 - j0.5)}{6 + j17.5} = \frac{6 + j17.5}{6 + j17.5} = 1$$

$$Z_2 = \frac{Z_AZ_C}{T} = \frac{(1 + j6)(2 + j12)}{6 + j17.5} = \frac{-70 + j24}{6 + j17.5}$$

$$= \frac{(-70 + j24)(6 - j17.5)}{(6 + j17.5)(6 - j17.5)} = \frac{0 + j1369}{+342.25} = 0 + j4$$

$$Z_3 = \frac{Z_B Z_C}{T} = \frac{(3 - j0.5)(2 + j12)}{6 + j17.5} = \frac{12 + j35}{6 + j17.5} = 2$$

Similar mathematical transformations using different impedances may yield elements in the alternate circuit that are not physically achievable.

Example: $-1 + j4$ would have to consist of a negative resistance plus an impedance.

ELECTRICAL 6

Given the circuit and voltage measurements shown below:

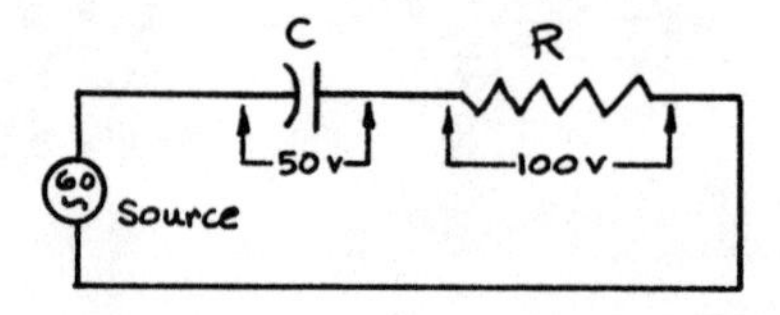

(a) What is the voltage of the 60 cycle source?
(b) What is the power factor of the circuit?

$IR = 100$ volts

$I(-jX_C) = 50$ volts

$$\therefore \bar{E} = IR - jIX_C = 100 - j50 \text{ volts}$$
(rectangular form)

$$\bar{E} = 112 \angle -26.6°$$

(a) $|E| = 112$ volts ●

(b) Power factor $= \cos 26.6° = 0.89$ (leading) ●

ELECTRICAL 7

Given the following combinations of voltage, frequency, and inductance, which circuit has the largest inductive reactance?

(a) 110 volts, 60 cps, 0.1 h
(b) 4000 volts, 60 cps, 1.0 h
(c) 110 volts, 10 kc, 0.1 h
(d) 4000 volts, 5 kc, 0.1 h
(e) 4000 volts, 5 kc, 0.01 h

Inductive reactance $X_L = 2\pi fL$

(a) $X_L = 2\pi(60)(0.1) = 37.7\ \Omega$

(b) $X_L = 2\pi(60)(1.0) = 377\ \Omega$

(c) $X_L = 2\pi(10 \times 10^3)(0.1) = 6280\ \Omega$ ●

(d) $X_L = 2\pi(5 \times 10^3)(0.1) = 3140\ \Omega$

(e) $X_L = 2\pi(5 \times 10^3)(0.01) = 314\ \Omega$

Answer is (c)

ELECTRICAL 8

In the bridge circuit shown below, no current is flowing through the ammeter. The value of R_x is

(a) 40 ohms
(b) 80 ohms
(c) 100 ohms
(d) 200 ohms
(e) 400 ohms

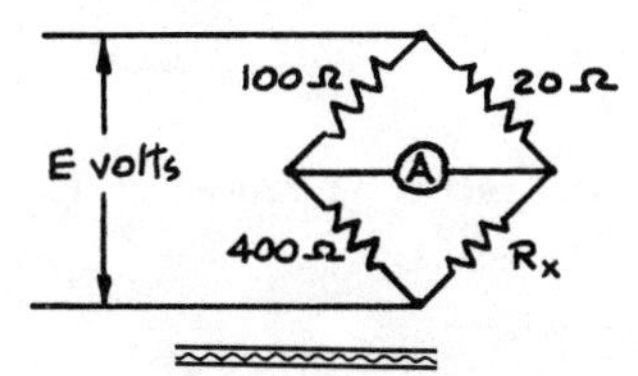

When there is no current through the ammeter the current through the 20 ohm resistance is equal to the current through R_x and the current current through the 100 ohm resistance is equal to the current through the 400 ohm resistance.

or $\quad I_{20} = I_{R_x}$ and $I_{100} = I_{400}$

The difference in potential $V_{100} = V_{20}$ and $V_{400} = V_{R_x}$

Then $\quad I_{20}R_{20} = I_{100}R_{100} \quad (1) \qquad$ and $\quad I_{20}R_x = I_{100}R_{400} \quad (2)$

dividing

$$\frac{(1)}{(2)} \qquad \frac{I_{20}R_{20}}{I_{20}R_x} = \frac{I_{100}R_{100}}{I_{100}R_{400}} \qquad \frac{R_{20}}{R_x} = \frac{R_{100}}{R_{400}}$$

$$R_x = \frac{R_{400}R_{20}}{R_{100}} = \frac{400(20)}{100} = 80 \text{ ohms}$$

Answer is (b)

ELECTRICAL 9

Two heating elements connected in series have resistances of 600 ohms and 300 ohms, and temperature coefficients α ($°C^{-1}$) of 0.001 and 0.004, respectively, at 20°C.

REQUIRED: (a) Find the resistance of the combination at a temperature of 50°C.

(b) What is the effective temperature coefficient of the combination?

(a) Assume the resistances are exactly 300 ohms and 600 ohms at 20°C and the temperature coefficients are positive.

The resistances will change according to the linear relation:

$$R_2 = R_1 + R_1\alpha(t_2 - t_1) \qquad (1)$$

For the 600 ohm element $R_{50°} = 600 + 600(0.001)(50 - 20) = 618$ ohms

For the 300 ohm element $R_{50°} = 300 + 300(0.004)(50 - 20) = 336$ ohms

The combined resistance = 954 ohms

(b) From equation (1) one sees that the temperature coefficient appears in the $R_1 \alpha(t_2 - t_1)$ term. Assuming $(t_2 - t_1)$ is the same for both heating elements, then the effective temperature coefficient is

$$\alpha_{eff} = \frac{R_{600}\,\alpha_{600} + R_{300}\,\alpha_{300}}{R_{600} + R_{300}} = \frac{600(0.001) + 300(0.004)}{900}$$

$$= \frac{0.6 + 1.2}{900} = 0.002$$ ●

ELECTRICAL 10

In the circuit shown below, the switch is closed at time t = 0.

REQUIRED: Find $e_o(t)$ at t = 1 second.

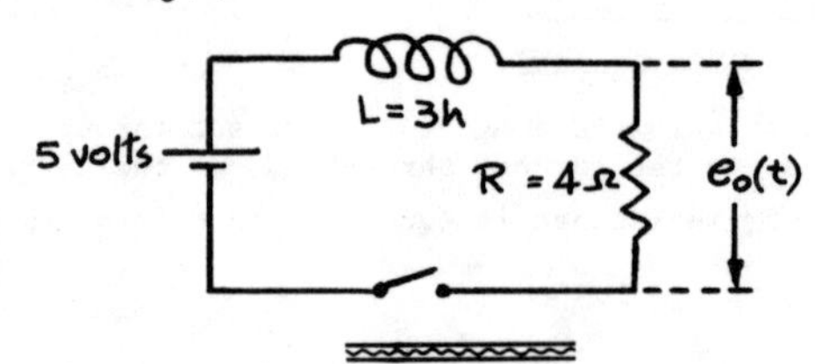

$$e_o(t) = i(t)4$$

The loop equation

$$V = L\frac{di}{dt} + Ri \qquad 5 = 3\frac{di}{dt} + 4i \qquad (1)$$

Assume the solution is $i(t) = Ae^{-t/\tau} + B$

and therefore $\frac{di}{dt} = -\frac{A}{\tau}e^{-t/\tau}$

Since the inductor prevents an instantaneous change in $i(t)$ $i(0) = 0$.

$$0 = A + B$$

At $t = \infty$ the circuit has achieved steady state and $i = \frac{V}{R} = 1.25$ amps

$$i(\infty) = Ae^{-\infty} + B = 1.25 \qquad i(\infty) = B = 1.25 \qquad A = -B = -1.25$$

Substituting into the assumed loop equation (1)

$$5 = 3\left[+\frac{1.25}{\tau}e^{-t/\tau}\right] + 4\left[-1.25\,e^{-t/\tau} + 1.25\right]$$

When t = 0 the equation still holds

$$5 = 3\left(\frac{1.25}{\tau}\right) \qquad \tau = \frac{3.75}{5} = 0.75 \text{ or } \frac{L}{R}$$

Now the output voltage

$$e_o(1) = (-1.25e^{-1/.75} + 1.25)4 = \left(\frac{-1.25}{3.78} + 1.25\right)4 = 3.68 \text{ volts}$$ ●

ELECTRICAL 11

Given the following DC circuit:

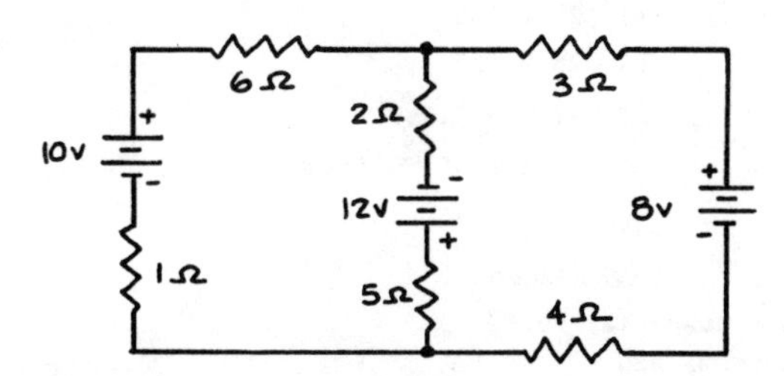

REQUIRED: Compute the current through the 12-volt battery, and the direction of current flow.

The circuit given may be converted into the following equivalent circuit.

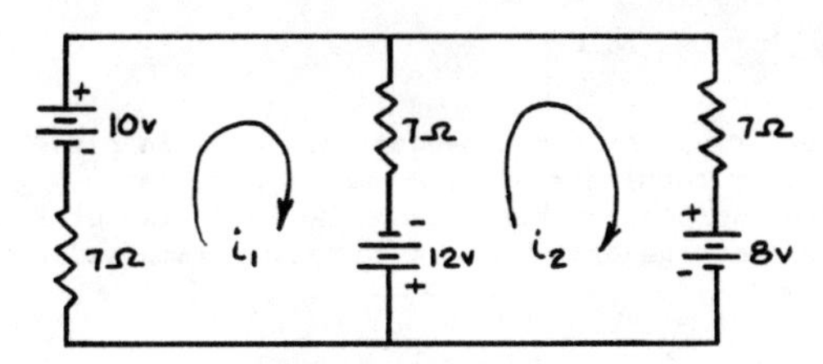

Use + current convention, and assume clockwise current direction in each mesh.

Write Kirchhoff equations for each mesh: $\sum E = \sum iR$

(1): $+10 + 12 = 14i_1 - 7i_2$

(2): $-12 - 8 = -7i_1 + 14i_2$

Solve the simultaneous equations for i_1 and i_2:

(3)=2x(1): $+44 = 28\,i_1 - 14i_2$

(3)+(2): $+24 = 21\,i_1$ $\quad \therefore i_1 = +1.143$ A (in assumed direction)

substitute i_1 in (1): $+22 = 14(+1.143) - 7i_2$

$i_2 = -0.857$ A (opposite assumed direction)

Current through 12v battery is $i_1 - i_2 = +1.143 - (-0.857) = +2.000$ A ●

The positive sign indicates current in the assumed direction from top to bottom of the circuit diagram.

ELECTRICAL 12

The polar form $25\angle{-45^\circ}$ is equivalent to which rectangular form of the complex number?

(a) $25 - j45$

(b) $25 - j25$

(c) $17.7 - j17.7$

(d) $25(\cos 45^\circ - j\sin 45^\circ)$

(e) $25\ e^{-j\pi/4}$

(c) is equivalent in rectangular form,
(d) in trigonometric form, and
(e) in exponential form.

Answer is (c) ●

ELECTRICAL 13

A circuit (two wire) supplies 600 incandescent lamps in parallel, with 110 volts between lamp terminals and 1/2 ampere per lamp. Neglect any voltage drop in that portion of circuit on the load side of first lamp, and assume that the voltage drop between the first lamp and the supply source is 2.2 volts.

Compute the
(a) losses in the supply line (watts)
(b) power required to serve the circuit
(c) voltage at the source
(d) the efficiency of the circuit

(a) Loss in the supply line = IE = 600 x 0.5 x 2.2 = 660 watts ●

(b) Power to serve the circuit P = IE = 600 x 0.5 x 112.2 = 33.66 Kw ●

(c) Voltage at the source = 110 + 2.2 = 112.2 volts ●

(d) Efficiency = output/input

$$= \frac{110}{112.2} \times 10^2 = 98\%$$ ●

ELECTRICAL 14

A voltmeter with an internal resistance of 5000 ohms is calibrated to read 10 volts at full scale. Show how this voltmeter may be modified to measure a voltage of 150 at full scale.

To increase the range of the voltmeter it is necessary to add a resistance in series.

Without the series resistance $I = \frac{E}{R} = \frac{10}{5000} = 0.002$ amp

To modify the voltage range $R_T = \frac{E}{I} = \frac{150}{0.002} = 75{,}000$ ohms

Required series resistance = 75,000 - 5,000 = 70,000 ohms

Modify the voltmeter by adding 70,000 ohms of resistance in series. ●

ELECTRICAL 15

Electrons normally flow

(a) From the grid to the plate in an electron tube
(b) From the + terminal to the - terminal inside a battery
(c) Uniformly distributed throughout the cross-section of a conductor in an AC circuit
(d) At the speed of light in a conductor
(e) Through a capacitor

Answer is (b) ●

ELECTRICAL 16

When an electric current is flowing in a conductor, heat is developed at a rate proportional to the square of the intensity of the current. The proportionality factor is the resistance of the conductor. This relation is known as

(a) Coulomb's Law
(b) Faraday's Law
(c) Joule's Law
(d) Kirchhoff's Rule
(e) Ohm's Law

Coulomb's Law

$$F = \frac{Q_1 Q_2}{K \epsilon d^2}$$

The force F between point charges Q_1 & Q_2 distant d with permittivity ϵ and dielectric constant K

Faraday's Laws of Electrolysis

1. The mass of a substance liberated in an electrolytic cell is proportional to the quantity of electricity passing through the cell.
2. When the same quantity of electricity is passed through different electrolytic cells, the masses of the substances liberated are proportional to their chemical equivalents.

One Faraday (96,500 coulombs) of electricity will liberate 1 gram equivalent weight of a substance.

Kirchhoff's Network Laws

1. In an electric circuit the sum of the currents flowing toward a junction is equal to the sum of the currents flowing away from the junction.
2. The algebraic sum of the emfs and the voltage drops around any closed path in an electric circuit is equal to zero.

Ohm's Law

$$I = \frac{E}{R}$$

Answer is (c) ●

ELECTRICAL 17

The result of capacitors C_1 = 9.6 microfarads and C_2 connected in series, is 7.3 microfarads. Capacitor C_2, in microfarads, is

(a) 2.3
(b) 30.5
(c) 35.0
(d) 49.3
(e) 84.5

For capacitors in series

$$\frac{1}{C_{eq}} = \frac{1}{C_1} + \frac{1}{C_2} \qquad \frac{1}{7.3} = \frac{1}{9.6} + \frac{1}{C_2}$$

$$0.1370 = 0.1042 + \frac{1}{C_2}$$

$$C_2 = \frac{1}{0.1370 - 0.1042} = \frac{1}{0.0328} = 30.5 \text{ microfarads}$$

Answer is (b) ●

ELECTRICAL 18

A resistor, R_1 = 20 ohms, is connected to a power source. A voltmeter of 10,000 ohms resistance reads 110 volts across R_1. Replacing R_1 with R_2 = 10 ohms, the voltmeter reads 100 volts.

Find the line resistance and the source voltage.

Case 1

$$\text{Current through } R_1 = \frac{E}{R} = \frac{110}{20} = 5.5 \text{ amps}$$

$$\text{Current through meter} = \frac{110}{10{,}000} = 0.01 \text{ amps}$$

$$\overline{5.51}$$

$$E = 5.51\, R_{line} + 110$$

Case 2

$$\text{Current through } R_2 = \frac{100}{10} = 10 \text{ amps}$$

$$\text{Current through meter} = \frac{100}{10{,}000} = 0.01 \text{ amps}$$

$$\overline{10.01}$$

$$E = 10.01\, R_{line} + 100$$

For constant source voltage

$$5.51\, R_{line} + 110 = 10.01\, R_{line} + 100$$

$$R_{line} = \frac{110 - 100}{10.01 - 5.51} = \frac{10}{4.5} = 2.22 \text{ ohms}$$ ●

Source voltage = 5.51(2.22) + 110 = 122.2 volts ●

ELECTRICAL 19

Design a circuit whose output voltage increases with time and passes through 40 volts 1.5 seconds after a switch is operated.
An ungrounded, 100-volt, direct current low-impedance source is available.

Describe the operation and include provision for resetting the circuit in your design.

$$\frac{q}{q_{max}} = 1 - e^{-\frac{t}{RC}}$$

$$\frac{40}{100} = 1 - e^{-\frac{1.5}{RC}}$$

$$0.6 = e^{-\frac{1.5}{RC}}$$

$$\ln 0.6 = -\frac{1.5}{RC} = -0.510$$

$$RC = \frac{-1.5}{-0.510} = 2.94 \text{ seconds}$$

Many alternatives of R and C combinations are possible.
For instance, if C = 1 microfarad, then make R = 2.94 megohms. ●

Operation ●
Close Switch 1. Output voltage rises.
Open Switch 1 and close Switch 2 to discharge condenser. Open Switch 2.

ELECTRICAL 20

Twelve resistors, each one having a resistance as indicated on the diagram, are connected to form a cube as shown.

Find the total resistance between points A and B. Show calculations.

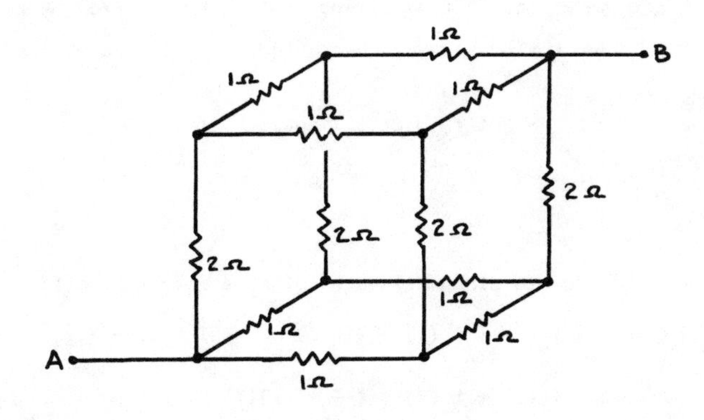

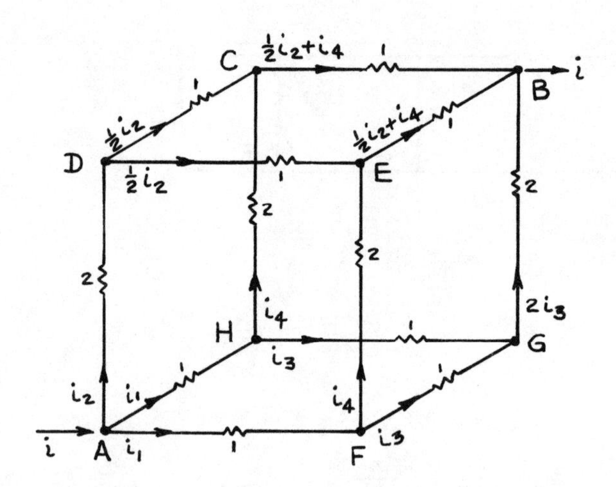

Label the nodes A through H
Apply a 1 volt test potential across the cube, from A to B.
If we can find the current i that flows, we can obtain the equivalent resistance R.

By symmetry we arrive at equipotential nodes

F = H and E = C

Since the potential difference from A to F and A to H is the same, and the resistances are the same, current i_1 flows in both legs AF and AH.
A different current i_2 flows in AD. By symmetry again we see that the current in DE and DC are equal and half of i_2 or $\frac{1}{2}i_2$. If current i_3 flows in FG, by symmetry i_3 also flows in HG. Let the current in FE be i_4. Then HC is also i_4. From EB and CB the current is $\frac{1}{2}i_2 + i_4$. From GB the current is $2i_3$. Thus the problem has been reduced to 5 unknown currents.

Apply Kirkhhoff's Laws.

1st Law Sum i_{in} = Sum i_{out} at all nodes

At A $\quad i = 2i_1 + i_2 \qquad (1)$

At B $\quad i_1 = i_3 + i_4 \qquad (2)$

2nd Law E = Sum iR along all paths of same potential difference

Path AFGB $\quad 1 \text{ volt} = i_1(1) + i_3(1) + 2i_3(2) \qquad 1 = i_1 + 5i_3 \qquad (3)$

Path AFEB $\quad 1 \text{ volt} = i_1(1) + i_4(2) + (\frac{i_2}{2} + i_4)(1)$

$$1 = i_1 + 3i_4 + \frac{i_2}{2} \qquad (4)$$

Path ADEB $\quad 1 \text{ volt} = i_2(2) + \frac{i_2}{2}(1) + (\frac{i_2}{2} + i_4)(1)$

$$1 = 3i_2 + i_4 \qquad (5)$$

We now have 5 independent equations in 5 unknown currents. Solve for the currents.

Substituting the value of i_2 from (5) into (4)

$$1 = i_1 + 3i_4 + (\frac{1}{2})(\frac{1}{3})(-i_4 - 1) \qquad i_1 = \frac{5}{6} - \frac{17}{6} i_4 \qquad (6)$$

Substituting the value of i_3 from (3) into (2)

$$i_1 = \frac{1}{5} - \frac{1}{5} i_1 + i_4 \qquad \frac{6}{5} i_1 = i_4 + \frac{1}{5} \qquad (7)$$

Substituting the value of i_1 from (6) into (7)

$$\frac{6}{5}(\frac{5}{6} - \frac{17}{6} i_4) = i_4 + \frac{1}{5} \qquad 1 - \frac{17}{5} i_4 = i_4 + \frac{1}{5}$$

$$(\frac{5 + 17}{5}) i_4 = \frac{4}{5} \qquad i_4 = \frac{4}{22} = \frac{2}{11} \qquad (8)$$

Substituting (8) into (6)

$$i_1 = \frac{5}{6} - \frac{17}{6}(\frac{2}{11}) = \frac{55}{66} - \frac{34}{66} = \frac{21}{66} = \frac{7}{22} \qquad (9)$$

(8) into (5)

$$1 = 3i_2 + \frac{2}{11} \qquad i_2 = \frac{1}{3}(\frac{9}{11}) = \frac{9}{33} = \frac{6}{22} \qquad (10)$$

(9) into (3)

$$1 = \frac{7}{22} + 5i_3 \qquad i_3 = \frac{1}{5}(\frac{15}{22}) = \frac{3}{22} \qquad (11)$$

(9) and (10) into (1)

$$i = 2(\frac{7}{22}) + \frac{6}{22} = \frac{20}{22}$$

Now $R_{equiv} = \frac{E}{I} = \frac{1}{\frac{20}{22}} = \frac{22}{20} = 1.1$ ohm ●

ELECTRICAL 21

What is the coefficient of coupling of two coils whose mutual inductance is 1 Henry and whose self-inductances are 1.2 and 2 Henries?

(a) 0.417
(b) 0.646
(c) 1.549
(d) 2.041
(e) 2.400

$$k = \frac{M}{\sqrt{L_1 L_2}} = \frac{1}{\sqrt{(1.2)(2.0)}} = \frac{1}{1.55} = 0.645$$ ●

Answer is (b)

ELECTRICAL 22

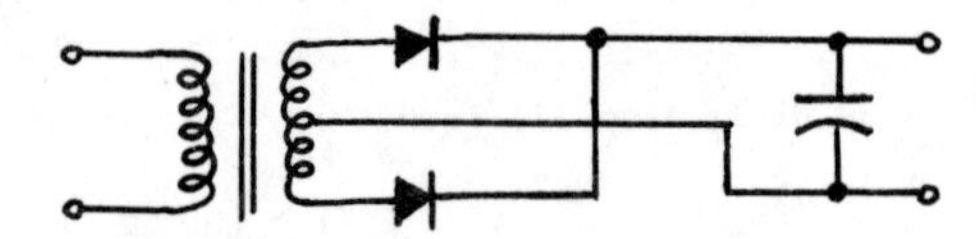

The circuit shown is a

(a) Full wave rectifier
(b) Push-Pull amplifier
(c) Oscillator
(d) Direct-coupled amplifier
(e) Bridge rectifier

The circuit is a full wave rectifier. One notes that by considering instantaneous polarities from the transformer output, rectification is obtained from each loop of the sine wave.

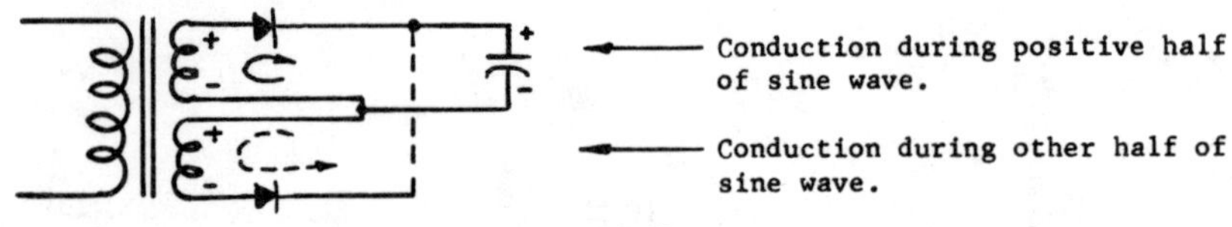

Answer is (a) ●

ELECTRICAL 23

The practical dimensional units of resistivity are:

(a) Ohms per circular mil foot
(b) Ohms - circular mil
(c) Ohm - circular mil per foot
(d) Circular mil - foot per ohm
(e) Ohm per foot

The resistance of a conductor

$R = \rho \frac{L}{A}$ where R = resistance in ohms
L = length in feet
A = area in circular mils
ρ = resistivity in ohm-circular mils/foot

Answer is (c) ●

ELECTRICAL 24

A "Henry" is the unit of

(a) Self or mutual inductance of a coil or coils
(b) The ratio of the capacitance to the resistance of a coil
(c) The ratio of the electrical charge to the resultant change of potential of a coil
(d) The total apparent resistance of a coil
(e) The natural frequency of a coil

$L = N \frac{d\phi}{di}$ = flux linkages per unit of current, or "henry" ●

Answer is (a)

ELECTRICAL 25

A shunt wound DC motor has an armature resistance of 3 ohms and a field resistance of 240 ohms. The terminal voltage is 120 VDC and the line current is 6 amps. The armature develops a back EMF when the motor is running.

Determine the mechanical power output and the efficiency of the motor, considering only the copper losses.

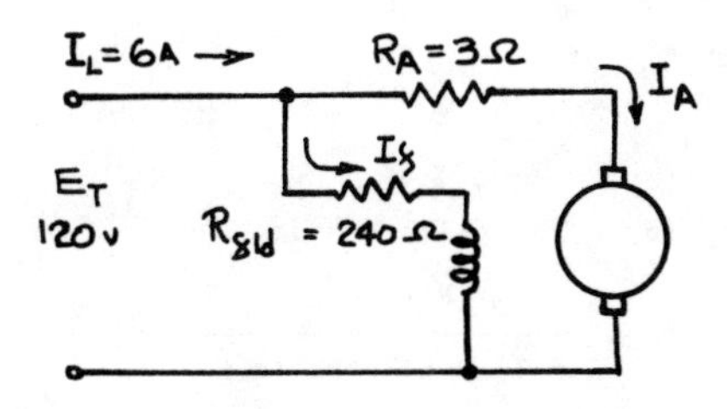

Total line current

$$I_L = 6\ A = I_f + I_A$$

where $I_f = \frac{E_T}{R_f} = \frac{120}{240} = 0.5\ A$

Therefore

$$I_A = 6 - 0.5 = 5.5\ A$$

Total voltage around armature circuit is given by Kirchhoff's Law:

$$E_T = R_A I_A + E_{b.emf} = 120\ v = 3(5.5) + E_{b.emf}$$

Therefore $E_{b.emf} = 120 - 16.5 = 103.5$ volts

$P_o = I_A E_{b.emf} = (103.5)(5.5) = 569$ watts ●

$$\text{Efficiency} = \frac{P_o}{P_{in}} = \frac{P_o}{P_o + \text{losses}} = \frac{P_o}{P_o + I_f^2 \cdot R_f + I_A^2 \cdot R_A}$$

$$= \frac{569}{569 + (0.5)^2(240) + (5.5)^2(3)} = \frac{569}{720.0}$$

$= 0.79 \cong 79\%$ ●

ELECTRICAL 26

A 100 KVA, 2400/240volt, 60 HZ, single phase transformer is subjected to short circuit and open circuit tests with the following results:

Short circuit primary voltage	= 48 volts
Total leakage resistance	= 0.65 ohms
Total leakage reactance	= 1.15 ohms
Open circuit power	= 240 watts

Compute the efficiency of the transformer at full load, 0.90 power factor lagging.

Since the test data gives the total leakage resistance and reactance, the test short circuit voltage of 48 volts is not needed.

The transformer equivalent circuit, in the simplest case, may be given as follows (here it is assumed that all values are referred to the primary side):

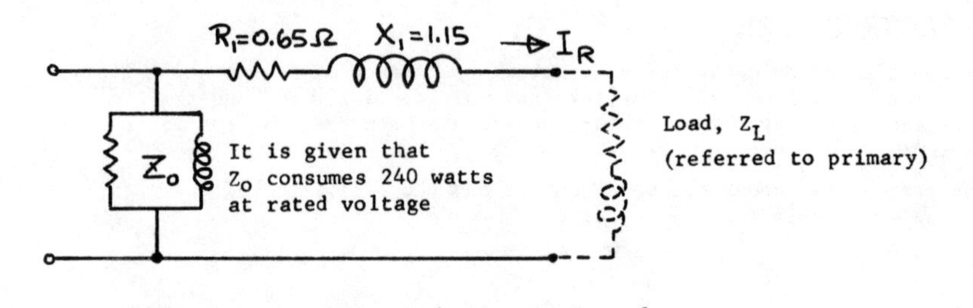

If one now assumes a rated current, I_R, of

$$I_R = \frac{100,000 \text{ VA}}{2,400 \text{ V}} = 41.6 \text{ A}$$

This is the load current, but referred to the primary side.

$$\text{Efficiency} = \frac{P_o}{P_{in}} = \frac{P_o}{P_o + \text{all losses}} = \frac{P_o}{P_o + I_R^2 \cdot R_1 + \text{loss in } Z_o}$$

where $P_o = 100,000 \cos \phi = 100,000(0.9) = 90,000$ watts

$$\text{Efficiency} = \frac{90,000}{90,000 + (41.6)^2(0.65) + 240} = \frac{90,000}{91,365} = 0.985 = 98.5\%$$ ●

ELECTRICAL 27

Under steady-state conditions a current "I" flows in a simple series circuit connected to a DC voltage source. To increase this steady-state current, you should increase the

(a) resistance
(b) voltage
(c) inductance
(d) capacitance
(e) power factor

Ohm's Law: $I = \frac{E}{R}$ To increase current, increase voltage. ●

Answer is (b)

ELECTRICAL 28

A 1500-watt heating element at an initial temperature of 15°C is connected to a supply voltage of 115 volts. After the temperature of the element has stabilized (steady - state conditions) it is found that a voltage of 125 volts is required to maintain rated output.

If the temperature coefficient of the element is 0.0006 per °C, determine the element's final temperature.

First, find R at t_1 when at 115 volts

$$R_1 = \frac{E^2}{P} = \frac{(115)^2}{1,500} = 8.82 \text{ ohms at } t_1 = 15°\text{C}$$

Then find R at t_2 when at 125 volts

$$R_2 = \frac{E^2}{P} = \frac{(125)^2}{1,500} = 10.4 \text{ ohms at } t_2 = ?$$

Then, since the temperature coefficient is given (and here one must assume that this coefficient is for $t_1 = 15^\circ C$), the basic temperature-resistance relationship is given as:

$$R_2 = R_1 \left[1 + \alpha_1 (t_2 - t_1) \right]$$

$$10.4 = 8.82(1 + 0.0006\,\Delta t)$$

$$0.0006\,\Delta t = 1.18 - 1 = 0.18 \qquad \Delta t = \frac{0.18}{0.0006} = 300^\circ C$$

Therefore $t_2 = 315^\circ C$ ●

ELECTRICAL 29

Find the resistance between points A and B in the figure shown below, if the value of each resistor is 2 ohms.

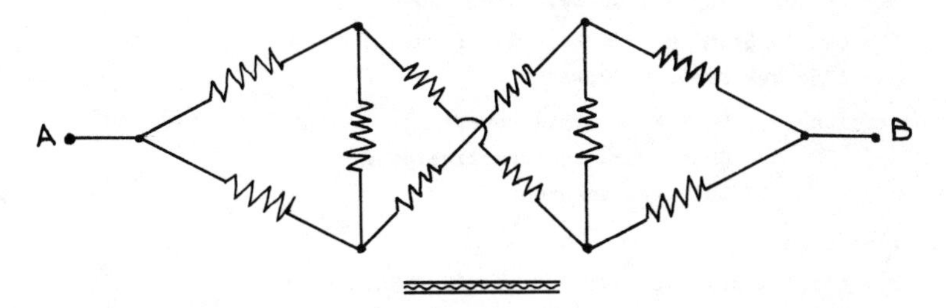

"Un-twisting" the network, and, since all resistors are equal, one immediately notes that there is no voltage difference across the vertical resistances, gives:

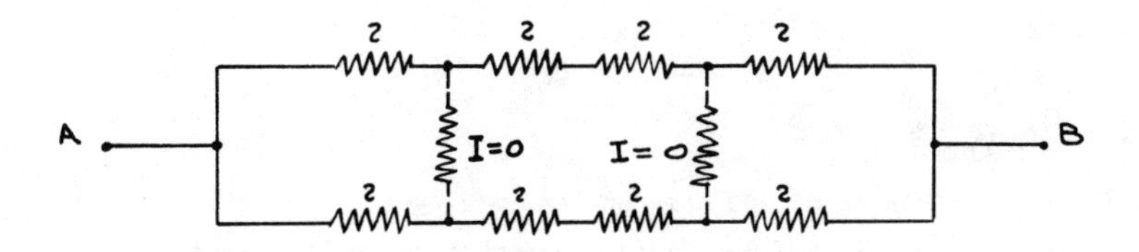

Thus R_{equiv} = 8 ohms in parallel with 8 ohms

$$= \frac{1}{\frac{1}{R_1} + \frac{1}{R_2}} = \frac{R_1 R_2}{R_1 + R_2} = \frac{(8)(8)}{8 + 8} = 4 \text{ ohms}$$ ●

ELECTRICAL 30

A synchronous motor is to be used to improve the power factor of a load of 1500 kilowatts at a power factor of 0.85 lagging.
Assume the synchronous motor operates at a power factor of 0.75 leading.
What must be the kva rating of the synchronous motor if it is required to raise the total power factor to 0.95 lagging?

Trig functions correstponding to the three power factors are tabulated:

pf = cos ϕ	sin ϕ	ϕ
0.85	0.527	31.8°
0.75	0.661	41.4°
0.95	0.312	18.2°

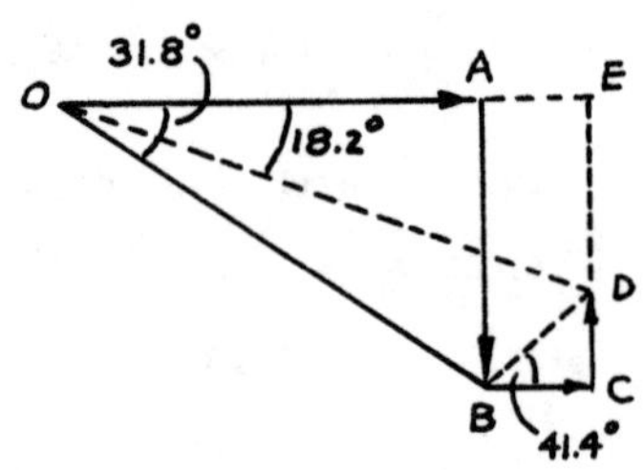

$OB = \frac{1500}{0.85} = 1765$ kva $\qquad AB = 1765(0.527) = 930$ kvar

The vector diagram shows the situation before correction:

$OA = OB \cos 31.8^\circ = 1500$ kw average power

$AB = OB \sin 31.8^\circ = 930$ kvar lag reactive power

$OB = 1765$ kva apparent power

The correction is: BC = kw average power

CD = kvar lead reactive power

BD = kva apparent power

After correction:

DE = total kvars lag = $AB - CD = 930 - BD(0.661)$

OE = total kw = $OA + BC = 1500 + BD(0.75)$

$\tan 18.2^\circ = \frac{0.312}{0.95} = 0.327$ which equals $\frac{DE}{OE}$

Solve for BD: $930 - 0.661\ BD = 0.327(1500 + 0.75\ BD)$

$0.905\ BD = 439.5$

$BD = 485$ kva ●

ELECTRICAL 31

A coil has a resistance of 100 ohms and an inductance of 10 millihenries.

What is the time domain voltage, v(t), that must be applied across the coil to produce a sinusoidal current of 20 milliamperes RMS through it at a frequency of 4000 Hz (cps)?

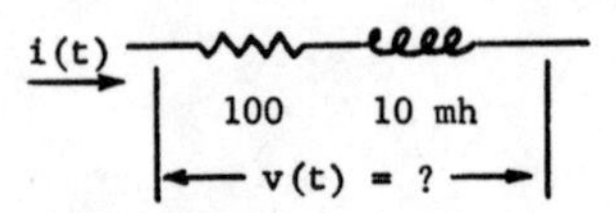

$\omega = 2\pi\,(4000) = 8000\pi$ rad/sec

$I_{max} = \sqrt{2}\,(0.020) = 0.0283 \qquad i(t) = I_{max} \sin \omega t = 0.0283 \sin(8000\pi t)$

$Z = R + j\omega L = 100 + j(8000\pi)(10 \times 10^{-3})$

$= 100 + j80\pi$

$|Z| = \sqrt{100^2 + (80\pi)^2} = 270$

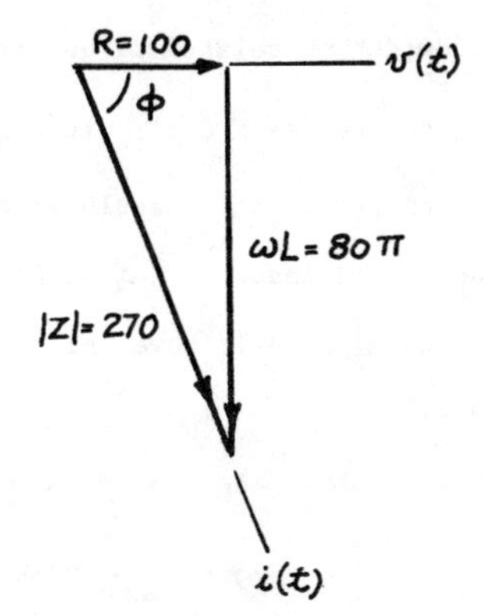

Current lags voltage (voltage leads current) by phase angle ϕ

$$\phi = \tan^{-1}\frac{80\pi}{100} = \tan^{-1} 2.51 = 68.3^o$$

$$v(t) = |Z|\, I_{max} \sin(\omega t + \phi)$$

$$= 270(0.0283)\sin(8000\pi t + 68.3^o)$$

$$= 7.64 \sin(25{,}100t + 68.3^o)$$ ●

ELECTRICAL 32

A circuit contains an inductance of 0.1062 mH, a capitance of 106 pf and a resistance in series. At resonance the impedance equals 10 ohms. Determine the resonant frequency and the band width at resonance (the frequency band between half-power points at resonance).

$L = 0.1062 \times 10^{-3}$ H, $C = 106 \times 10^{-12}$ farads

At resonance, current and voltage are both at maximum, in phase, and $X_L - X_C = 0$.

The circuit is purely resistive: $Z_o = R + j(X_L - X_C) = 10 + j0$

Since $Z_o = 10$ ohms at resonance, R= 10 ohms.

$X_L = \omega_o L$, and $X_C = \dfrac{1}{\omega_o C}$ where ω_o = rad/sec

Thus $\omega_o = \dfrac{1}{\sqrt{LC}} = \dfrac{1}{1.061 \times 10^{-7}} = 9.42 \times 10^6$ rad/sec at resonance

$f_o = \dfrac{\omega_o}{2\pi} = \dfrac{9.42 \times 10^6}{6.28} = 1.5 \times 10^6$ cycles/sec = 1.5 mHz ● resonant frequency

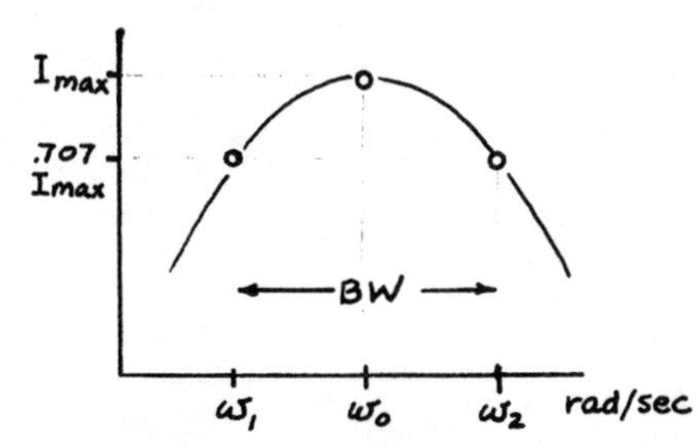

Lower ω_1 and higher ω_2 half power points exist where current leads 45° and lags 45°, respectively.

Bandwidth (BW) $= f_2 - f_1 = \dfrac{\omega_2 - \omega_1}{2\pi}$

For the series RLC circuit: $BW = \dfrac{R}{2\pi L} = \dfrac{10}{6.28(0.1062 \times 10^{-3})}$

$= 1.5 \times 10^4$ cycles/sec = 15 kHz ●

Bandwidth is related to quality factor Q by $Q = \frac{f_o}{BW}$.

For the series RLC circuit: $Q = \frac{\omega_o L}{R} = \frac{1}{\omega_o CR}$ (Q = 100 in this problem)

f_1 and f_2 are not exactly located at $f_o \pm \frac{BW}{2}$ although this often is a good approximation. ω_o is the geometric mean of ω_1 and ω_2. $\omega_o = \sqrt{\omega_1 \omega_2}$

Since $|Z|$ at half power points $= \sqrt{2}\,|Z_o|$, exact values of ω_1 and ω_2 may be calculated.

At lower ω_1: $X_L - X_C = -R$

$$\omega_1 L - \frac{1}{\omega_1 C} = -R$$

(ω_1 can be determined by rearrangement, substitution of R, L and C values, and solution by quadratic formula.)

At higher ω_2: $X_L - X_C = +R$

$$\omega_2 L - \frac{1}{\omega_2 C} = +R$$

(ω_2 can be similarly determined.)

9

Engineering Economics

Engineering economics (often called engineering economic analysis) is a group of techniques for the systematic analysis of alternative courses of action. This chapter begins with an introduction to engineering economics fundamentals, followed by problems and solutions.

Engineering economics problems are typically solved by use of a set of compound interest tables. While hand held calculators can solve many of the problems in this chapter quickly, they cannot yet work all of them. Because space prevents their publication here, readers must refer to a set of tables published elsewhere. One of the most complete sets will be found in Newnan: *Engineering Economic Analysis* (Engineering Press, Inc., P.O. Box 1, San Jose, CA 95103). A somewhat smaller, but probably adequate, set is contained in Newnan and Larock: *Engineering Fundamentals Examination Review*.

CASH FLOW

In examining alternative ways of solving a problem we recognize the need to resolve the various consequences (both favorable and unfavorable) of each alternative into some common unit. One convenient unit - and the one typically used in engineering economics - is money. Thus an initial step in resolving these problems is to convert the various consequences of an alternative into a table of year-by-year cash flows.

For example, a simple problem might be to portray the consequences of purchasing a new car as follows.

	Year	Cash Flow	
Beginning of first year	0	-4500	Car purchased 'now' for $4500 cash. The minus sign indicates a disbursement.
End of year	1	-350	Maintenance costs are $350 per year.
End of year	2	-350	
End of year	3	-350	
End of year	4	-350	
		+2000	The car is sold at the end of the 4th year for $2000. The plus sign represents a receipt of money.

This cash flow is represented graphically on the next page.

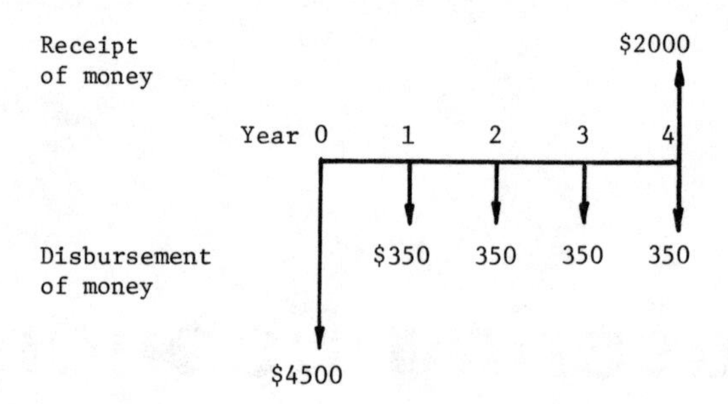

The upward arrow represents a receipt of money, and the downward arrows represent disbursements. The *x*-axis represents the passage of time.

TIME VALUE OF MONEY

When the money consequences of an alternative occur in a short period of time - say less than one year - we might simply algebraically add up the various sums of money and obtain the net result. But we cannot treat money this same way over longer periods of time. This is because money today is not the same as money at some future time.

Consider this question. Which would you prefer, $100 today or the assurance of receiving $100 a year from now? Clearly you would prefer the $100 today. If you had the money today, rather than a year from now, you could use it for the year. And if you had no use for it, you could lend it to someone who would pay interest for the privilege of using your money for the year. Thus $100 today would be equivalent, at 10% interest, to $110 a year hence.

EQUIVALENCE

We see that money at different points in time may be equal in the sense that it is $100, but $100 a year hence is not an acceptable substitute for $100 today. When we have acceptable substitutes we say they are *equivalent* to each other. Thus at 10% interest, $110 a year hence is equivalent to $100 now.

Equivalence is an essential factor in engineering economics. Suppose we wish to select the better of two alternatives. First, we must compute their cash flows. An example is:

	Alternative	
Year	*A*	*B*
0	-$2000	-$2800
1	+800	+1100
2	+800	+1100
3	+800	+1100

The larger investment in Alternative *B* results in larger subsequent benefits, but we have no direct way of knowing if Alternative *B* is better than Alternative *A*. Therefore, we do not know which alternative should be selected. To make a decision we must resolve the alternatives into *equivalent* sums so they may be compared accurately and a decision made.

COMPOUND INTEREST

To facilitate equivalence computations a series of compound interest factors are used.

Symbols

Present sum .... P

Future sum .... F

End-of-period payments or receipts in a uniform series, continuing for a specified number of periods .... A

Number of interest periods .... n

Interest rate per interest period i

Single Payment Formulas

Suppose a present sum of money P is invested for one year at interest rate i. At the end of the year we would receive back our initial investment P together with interest equal to Pi, or a total amount P+Pi. Factoring P, the sum as the end of one year is P(1+i). If we agree to let our investment remain for subsequent years, the progression is as follows.

	Amount at beginning of period	+	Interest for the period	=	Amount at end of the period
1st year	P		Pi		$P(1+i)$
2nd year	$P(1+i)$		$Pi(1+i)$		$P(1+i)^2$
3rd year	$P(1+i)^2$		$Pi(1+i)^2$		$P(1+i)^3$
nth year	$P(1+i)^{n-1}$		$Pi(1+i)^{n-1}$		$P(1+i)^n$

The present sum P increases in n periods to $P(1+i)^n$. This gives us a relationship between a present sum P and its equivalent future sum F.

$$\text{Future sum} = (\text{Present sum})(1+i)^n$$

$$F = P(1+i)^n$$

This is the *single payment compound amount factor*. In functional notation it is written:

$$F = P(F/P,i\%,n)$$

The relationship may be rewritten as

$$\text{Present sum} = (\text{Future sum})(1+i)^{-n}$$

$$P = F(1+i)^{-n}$$

This is the *single payment present worth factor*. It is written

$$P = F(P/F,i\%,n)$$

Uniform Series Formulas

Consider the following situation.

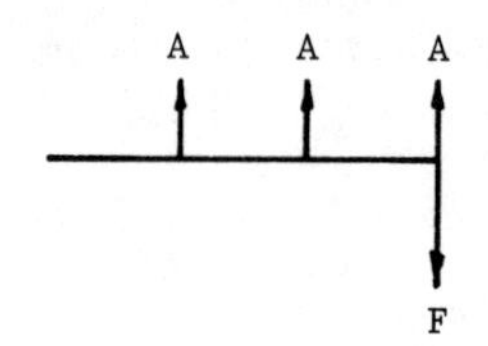

A = end-of-period receipt or disbursement in a uniform series, continuing for n periods.

F = a future sum of money.

Using the single payment compound amount factor, we can write an equation for F in terms of A.

$$F = A + A(1+i) + A(1+i)^2 \tag{1}$$

In our situation, with n = 3, Eqn 1 may be written in a more general form.

$$F = A + A(1+i) + A(1+i)^{n-1} \tag{2}$$

Multiply Eqn 2 by (1+i)

$$(1+i)F = A(1+i) + A(1+i)^{n-1} + A(1+i)^n \tag{3}$$

Write Eqn 2

$$F = A + A(1+i) + A(1+i)^{n-1} \tag{2}$$

Eqn 3 - Eqn 2

$$iF = -A + A(1+i)^n$$

$$F = A\left[\frac{(1+i)^n - 1}{i}\right]$$ *Uniform Series Compound Amount Factor*

Solving this equation for A

$$A = F\left[\frac{i}{(1+i)^n - 1}\right]$$ *Uniform Series Sinking Fund Factor*

Since $F = P(1+i)^n$, we can substitute this expression for F in the equation and obtain

$$A = P\left[\frac{i(1+i)^n}{(1+i)^n - 1}\right]$$ *Uniform Series Capital Recovery Factor*

Solving this equation for P

$$P = A\left[\frac{(1+i)^n - 1}{i(1+i)^n}\right]$$ *Uniform Series Present Worth Factor*

In functional notation the uniform series factors are:

Compound Amount	(F/A,i%,n)
Sinking Fund	(A/F,i%,n)
Capital Recovery	(A/P,i%,n)
Present Worth	(P/A,i%,n)

PRESENT WORTH

With the compound interest factors one may alter a cash flow into an equivalent sum or an equivalent cash flow. The three principal methods for comparing alternatives are Present Worth, Annual Cost, and Rate of Return. In this section we begin with Present Worth

Criteria

Engineering economics problems inevitably fall into one of three categories.

1. Fixed Input — The amount of money or other input resources is fixed.
   Example: A project engineer has a budget of $450,000 to overhaul a petroleum process unit.

2. Fixed Output — There is a fixed task, or other output, to be accomplished.
   Example: A mechanical contractor has been awarded a fixed price contract to air condition a building.

3. Neither Input nor Output Fixed — This is the general situation where neither the amount of money or other inputs, nor the amount of benefits or other outputs are fixed.
   Example: A consulting firm has more work available than it can handle. It is considering paying the staff for working evenings to increase the amount of design work it can perform.

In each of the categories we can determine what the criterion should be for economic efficiency. For Present Worth analysis the proper criteria are:

Category	Present Worth Criterion
Fixed Input	Maximize the present worth of benefits or other outputs.
Fixed Output	Minimize the present worth of costs or other inputs.
Neither Input nor Output Fixed	Maximize [present worth of benefits minus present worth of costs] or stated another way, Maximize Net Present Worth.

Application of Present Worth Analysis

Present worth analysis is frequently used to determine the present value of future money receipts and disbursements. We might want to know, for example, the present worth of an income producing property, like an oil well. This should provide an estimate of the price at which the property could be bought or sold.

An important restriction in the use of present worth calculations is that there must be a common analysis period when comparing alternatives. It would be incorrect, for example, to compare the present worth (PW) of cost of Pump A, expected to last six years, with the PW of cost of Pump B, which is expected to last twelve years.

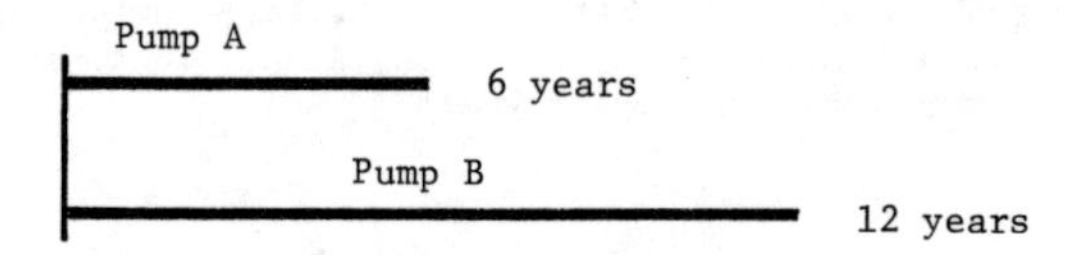

Improper Present Worth Comparison

In situations like this the solution is either to use some other analysis technique (generally the annual cost method) or to restructure the problem so there is a common analysis period. In the example above, a customary assumption would be that there is a need for the pump for twelve years, and that Pump A will be replaced by an identical Pump A at the end of six years. This gives a 12-year common analysis period.

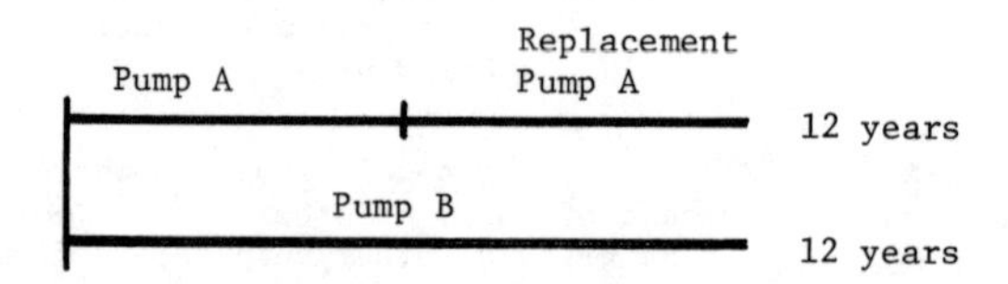

Correct Present Worth Comparison

This approach is easy to use when the different lives of the alternatives have a practical least common multiple life. When this is not true, for example, life of X equals 7 years, and life of Y equals 11 years, some assumptions must be made, or the present worth method should not be used.

Capitalized Cost

In the special situation where the analysis period is infinite ($n = \infty$), an analysis of the present worth of cost is called *capitalized cost*. There are relatively few situations where the analysis period is infinity. It does occur in some public projects, permanent endowments, and cemetery perpetual care.

When n equals infinity, a present sum P will accrue interest of Pi for every future interest period. For the principal sum P to continue undiminished (an essential requirement when n equals infinity), the end-of-period sum A that can be disbursed is Pi.

$$P \longrightarrow P + Pi \longrightarrow P + Pi \longrightarrow P + Pi \longrightarrow \cdots$$

(each $P + Pi$ with an arrow down to A)

When $n = \infty$ the fundamental relationship between P, A, and i is

$$A = Pi$$

Some form of this equation is used whenever there is a problem with an infinite analysis period.

FUTURE WORTH

In present worth analysis the comparison is made in terms of the equivalent *present* costs and benefits. The selection of the present as the point for the computations is simply a matter of convenience. The computations may be made at any point in time. When the future results are desired, we compute the equivalent *future* costs and benefits and call the computations Future Worth analysis.

ANNUAL COST

The annual cost method is more accurately described as the method of Equivalent Uniform Annual Cost (EUAC) or where the computation is of benefits, the method of Equivalent Uniform Annual Benefits (EUAB).

Criteria

For each of the three possible categories of problems there are annual cost criteria for economic efficiency.

Category	Annual Cost Criterion
Fixed Input	Maximize the equivalent uniform annual benefits, that is, Maximize EUAB.
Fixed Output	Minimize the equivalent uniform annual cost, that is, Minimize EUAC.
Neither Input nor Output Fixed	Maximize (EUAB - EUAC).

Application of Annual Cost Method

In the present worth section we saw that the present worth method requires that there be a common analysis period for all alternatives. This same restriction does not apply in all annual cost calculations, but it is important to understand the circumstances that justify comparing alternatives with different useful lives.

Often an analysis is to provide for a more or less continuing requirement. One might need to pump water from a well, for example, as a continuing requirement. Regardless of whether the pump has a useful life of 6 years or 12 years, we would select the one whose annual cost is a minimum. This would still be the case if the pump useful lives were the more troublesome 7 and 11 years, respectively. Thus if we can assume a continuing need for an item, an annual cost comparison among alternatives of different useful service lives is valid. Frequently, examination problems specify solution by the annual cost method and there is no information concerning the appropriate analysis period. In this situation, the assumption of a continuing requirement appears reasonable.

The underlying assumption in these situations is that when the shorter lived alternative has reached the end of its useful service life, it can be replaced with an identical item with identical costs and so forth. Therefore the EUAC of the initial alternative for a relatively short period of time is equal to the EUAC for the continuing series of replacements.

If, on the other hand, there is a specific requirement in some situation to pump water for ten years, then each pump must be evaluated to see what costs will be incurred during the analysis period and what salvage value, if any, may be recovered at the end of the analysis period. The annual cost analysis would need to consider the actual circumstances of the situation.

RATE OF RETURN

In the rate of return method the typical situation is where there is a cash flow representing the costs and benefits. The rate of return may be defined as the interest rate where

Present Worth of Cost = Present Worth of Benefits

Equivalent Uniform Annual Cost = Equivalent Uniform Annual Benefits

or Present Worth of Cost - Present Worth of Benefits = 0

These calculations frequently require trial and error solutions.

Rate of Return Analysis

Criterion for Two Alternatives

Compute the incremental rate of return on the cash flow representing the difference between the higher cost alternative and the lower cost alternative. If this rate of return is greater than or equal to the predetermined minimum attractive rate of return (or interest rate)* choose the higher cost alternative; otherwise choose the lower cost alternative.

EXAMPLE.

On the second page of this chapter two alternatives were described in terms of cash flows:

	Alternative	
Year	*A*	*B*
0	-\$2000	-\$2800
1	+800	+1100
2	+800	+1100
3	+800	+1100

If 5% is considered the minimum attractive rate of return, which alternative should be selected?

Solution

The computation of the rate of return for each alternative shows that Alternative *A* has a 9.7% rate of return, and Alternative *B* has an 8.7% rate of return. Thus both alternatives have rates of return in excess of the 5% minimum attractive rate of return.

The conventional assumption in economic analysis is that we will select the larger investment (rather than the smaller one) if the rate of return on the *additional investment* is greater than or equal to the minimum attractive rate of return. This means there will be situations where the alternative selected is *not* the alternative with the largest rate of return.

To decide which alternative in this example to select, compute the cash flow that represents the difference between the alternatives. Then compute the rate of return on the difference between the alternatives.

*The minimum attractive rate of return is often abbreviated as MARR.

Year	Alternative A	B	Difference between alternatives B - A
0	-$2000	-$2800	-$800
1	+800	+1100	+300
2	+800	+1100	+300
3	+800	+1100	+300
Rate of Return	9.7%	8.7%	6.1%

Since the rate of return on the difference between the alternatives, 6.1%, exceeds the 5% MARR, the increment of additional investment is desirable. In this example we would select Alternative B.

When there are three or more mutually exclusive alternatives, one must proceed following the same general logic presented for two alternatives. The components of incremental analysis are:

1. Compute the rate of return for each alternative. Reject any alternative where the rate of return is less than the given MARR. [This step is not not essential, but helps to immediately eliminate unacceptable alternatives.] One must insure that the lowest cost alternative has a rate of return greater than or equal to the MARR.

2. Rank the remaining alternatives in their order of increasing cost.

3. Examine the difference between the two lowest cost alternatives, as described for the two alternative problem. Select the better of the two alternatives, and reject the other one.

4. Take the preferred alternative from Step 3. Add the next higher cost alternative and proceed with another two-alternative comparison.

5. Continue until all alternatives have been examined and the best of the multiple alternatives has been identified.

BREAKEVEN ANALYSIS

Breakeven is the point where two alternatives are equivalent. This method is used to find the value of a single variable, with all other values held fixed. This may be described by an example.

	Alternative A	B
Initial cost	$500	P
Annual maintenance	$100	$50
Useful life	5 yrs	5 yrs
End of useful life salvage value, S	$150	$250

Interest rate: 8%

At the breakeven point how much could one afford to pay for Alternative B?

$$\text{EUAC of Alt. A} = 500(A/P,8\%,5) + 100 - 150(A/F,8\%,5) = \$199.68$$

$$\text{EUAC of Alt. B} = P(A/P,8\%,5) + 50 - 250(A/F,8\%,5)$$
$$= 0.2505P + 50 - 42.63$$

At breakeven: $EUAC_A = EUAC_B$

$$P = (199.68 - 50 + 42.63)/(0.2505) = \$767.70$$

VALUATION AND DEPRECIATION

One method of valuation of capital assets is called *Book Value* by accountants. Book Value is the cost of the asset, minus depreciation taken to that point in time. Thus, to compute book value one must understand the computation of depreciation.

Depreciation is defined, in its accounting sense, as the systematic allocation of the cost of a capital asset over its useful life. In computing a schedule of depreciation charges, three items are considered.

1. Cost of the property, P.
2. Useful life in years, n.
3. Salvage value of the property at the end of its useful life, S.

Three methods of depreciation are:

Straight Line Depreciation

$$\text{Depreciation charge in any year} = \frac{P - S}{n}$$

Sum-Of-Years Digits Depreciation

$$\text{Depreciation charge in any year} = \frac{\text{Remaining useful life at beginning of year}}{\text{Sum-Of-Years Digits for total useful life}}(P - S)$$

$$\text{where Sum-Of-Years Digits} = 1 + 2 + 3 + \ldots + n = \frac{n}{2}(n + 1)$$

Double Declining Balance Depreciation

$$\text{Depreciation charge in any year} = \frac{2}{n}(P - \text{Depreciation charges to date})$$

ENGR ECON 1

About how long will it take for $10,000 invested at 5% per year, compounded annually, to double in value?

(a) 5 yrs
(b) 10 yrs
(c) 15 yrs
(d) 20 yrs
(e) 25 yrs

P = $10,000 F = $20,000 i = 0.05 n = unknown

Using the single payment compound amount factor

$F = P(1+i)^n$ $1.05^n = \frac{20,000}{10,000} = 2$ n = 14.2 yrs ●

Alternate solution using compound interest tables

F = P(F/P,5%,n) (F/P,5%,n) = 20,000/10,000 = 2

From 5% interest tables: (F/P,5%,14) = 1.98 (F/P,5%,15) = 2.08

n = 14.2 ●

Answer is (c)

ENGR ECON 2

If \$200 is deposited in a savings account at the beginning of each of 15 years and the account draws interest at 7% per year, compounded annually, the value of the account at the end of 15 years will be most nearly

(a) \$5000
(b) 5400
(c) 6000
(d) 6900
(e) 7200

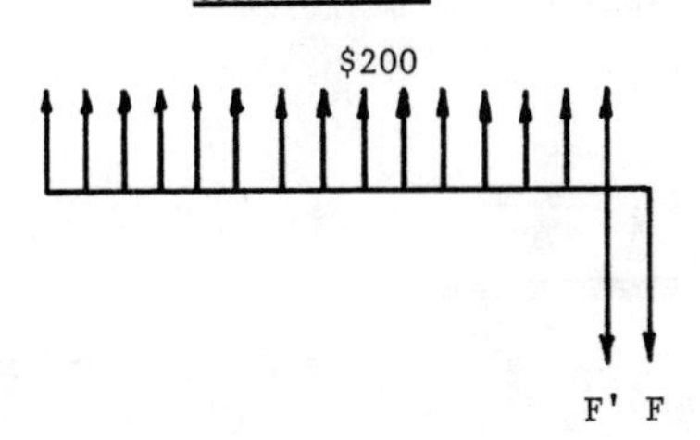

$F' = A(F/A,i\%,n) = \$200(F/A,7\%,15) = 200(25.129) = \$5025.80$

$F = F'(F/P,i\%,n) = 5025.80(F/P,7\%,1) = 5025.80(1.07) = \$5377.61$ ●

Answer is (b)

ENGR ECON 3

How many months at an interest rate of 1 percent per month does money have to be invested before it will double in value?

(a) 59 months
(b) 62
(c) 70
(d) 76
(e) 83

Let $P = \$1$ $F = \$2$ $i = 0.01$ per month n = number of months

$F = P(1 + i)^n$ $\$2 = \$1(1.01)^n$ $1.01^n = 2$

The solution may be obtained from a 1% compound interest table or by hand calculator. $n = 70$ months. ●

Answer is (c)

ENGR ECON 4

A department store charges one and one-half percent interest per month on credit purchases. This is equivalent to a nominal annual interest rate of

(a) 1.5 percent
(b) 15.0
(c) 18.0
(d) 19.6
(e) 21.0

The nominal interest rate is the annual interest rate ignoring the effect of any compounding.

Nominal interest rate = $1\frac{1}{2}$ percent/month x 12 months = 18% ●

Answer is (c)

ENGR ECON 5

A bank charges $1\frac{1}{2}$% per month on the unpaid balance for purchases made with its credit card. This is equivalent to what effective annual interest rate?

(a) 1.5%
(b) 12%
(c) 18%
(d) 19.5%
(e) 39%

Effective interest rate is the annual interest rate, taking into account the effect of any compounding during the year.

Effective interest rate = $(1 + i)^m - 1$ where i = interest rate/interest period.
m = number of compoundings per year.

$= (1 + 0.015)^{12} - 1$
$= 0.1956 = 19.56\%$ ●

Answer is (d)

ENGR ECON 6

A bank pays one percent interest on savings accounts four times a year. The effective annual interest rate is

(a) 1.00%
(b) 1.04%
(c) 3.96%
(d) 4.00%
(e) 4.06%

Effective interest rate = $(1 + 0.01)^4 - 1 = 0.0406 = 4.06\%$ ●

Answer is (e)

ENGR ECON 7

What interest rate, compounded quarterly, is equivalent to a 9.31% effective interest rate?

(a) 2.25%
(b) 2.33%
(c) 4.66%
(d) 9.00%
(e) 9.31%

$0.0931 = (1 + i)^4 - 1$

$1.0931^{0.25} = 1 + i$

$1.0225 = 1 + i$

i = 2.25% per quarter
i = 9% annual interest ●

Answer is (d)

ENGR ECON 8

A principal sum P is invested at a nominal interest rate r, compounded m times a year, for n years. The accumulated amount at the end of this period will be

(a) $P(1 + r/m)^r$
(b) $P(1 + r/m)^m$
(c) $P(1 + r/m)^{n/m}$
(d) $P(1 + r/m)^{m/r}$
(e) None of these

The relationship between a future sum F and a present sum P is

$F = P(1 + i)^n$ where i = interest rate per interest period
n = number of interest periods

The nominal annual interest rate r is the sum of the individual interest payments in a one year period. Therefore, the interest rate per interest period $i = r/m$. The number of interest periods in a particular situation is the number of interest periods per year m times the number of years n. Note that the problem specifies n years, while conventional notation defines n as the number of interest periods. Thus the problem makes the number of interest periods equal to the number of compoundings per year m times the number of years.

$i = r/m$ and no. of interest periods = mn

Substituting these into

$F = P(1 + i)^n$ gives $F = P(1 + r/m)^{mn}$ ●

Answer is (e)

ENGR ECON 9

In the formula $P = F(1 + i)^{-n}$ the factor $(1 + i)^{-n}$ is called the

(a) sinking fund factor
(b) single payment present worth factor
(c) single payment compound amount factor
(d) capital recovery factor
(e) uniform series present worth factor

Answer is (b) ●

ENGR ECON 10

A fund established to produce a desired amount at the end of a given period by means of a series of payments throughout the period is called a sinking fund, and is represented by the formula:

(a) $A = F\left[\frac{i}{(1 + i)^n - 1}\right]$

(b) $A = F\left[\frac{(1 + i)^n}{i}\right]$

(c) $A = P(1 + i)^{-n}$

(d) $A = P\left[\frac{(1 + i)^n - 1}{i(1 + i)^n}\right]$

(e) $A = F\left[\frac{i}{(1 + i)^n}\right]$

$$A = F\left[\frac{i}{(1+i)^n - 1}\right] \bullet$$

Answer is (a)

ENGR ECON 11

Which of the following relationships between compound interest factors is NOT correct?

(a) Single payment compound amount factor and single payment present worth factor are reciprocals.

(b) Sinking fund factor and uniform series compound amount factor are reciprocals.

(c) Capital recovery factor and uniform series present worth factor are reciprocals.

(d) Capital recovery factor equals sinking fund factor plus the interest rate.

(e) Capital recovery factor and sinking fund factor are reciprocals.

Since answer (c) says Capital recovery = 1/(Series present worth)
and (e) says Capital recovery = 1/(Sinking fund)
it would necessarily follow that Series present worth = Sinking fund, and that obviously is not true. Thus the incorrect statement is either (c) or (e).

(a) $(F/P,i\%,n) = \frac{1}{(P/F,i\%,n)}$

(b) $(A/F,i\%,n) = \frac{1}{(F/A,i\%,n)}$

(c) $(A/P,i\%,n) = \frac{1}{(P/A,i\%,n)}$

(d) $(A/P,i\%,n) = (A/F,i\%,n) + i$

(e) $(A/P,i\%,n) \neq \frac{1}{(A/F,i\%,n)}$ ●

Answer is (e)

ENGR ECON 12

For some interest rate i and some number of interest periods n, the uniform series capital recovery factor is 0.0854 and the sinking fund factor is 0.0404. The interest rate i must be

(a) $3\frac{1}{2}\%$

(b) $4\frac{1}{2}\%$

(c) 6%

(d) 8%

(e) 9%

In Part (d) of ENGR ECON 11 we saw that

$$(A/P,i\%,n) = (A/F,i\%,n) + i$$

If we substitute the values given here into the equation, we have

$$0.0854 = 0.0404 + i$$
$$i = 0.0854 - 0.0404 = 0.045 = 4\tfrac{1}{2}\% \; ●$$

Answer is (b)

ENGR ECON 13

An "annuity" is defined as

(a) Earned interest due at the end of each interest period.
(b) Cost of producing a product or rendering a service.
(c) Total annual overhead assigned to a unit of production.
(d) Amount of interest earned by a unit of principal in a unit of time.
(e) A series of equal payments occurring at equal periods of time.

Answer is (e) ●

ENGR ECON 14

An individual wishes to deposit a certain quantity of money now so that at the end of five years he will have \$500. With interest at 4% per year, compounded semiannually, how much must he deposit now?

(a) \$340.30
(b) 400.00
(c) 410.15
(d) 416.95
(e) 608.35

$P = F(1 + i)^{-n}$ where F = \$500.
i = 0.02 per interest period (i is *not* 0.04)
n = 10 interest periods

Calculator solution:

$$P = \$500(1 + 0.02)^{-10} = \$410.17 \; ●$$

Compound interest table solution:

$$P = \$500(P/F,2\%,10) = \$500(0.8203) = \$410.15 \; ●$$

Answer is (c)

ENGR ECON 15

The present worth of an obligation of \$10,000 due in 10 years if money is worth 9% is nearest to

(a) \$10,000
(b) 9,000
(c) 7,500
(d) 6,000
(e) 4,500

$P = F(P/F,9\%,10) = \$10,000(0.4224) = \$4224$ ●

Answer is (e)

ENGR ECON 16

$1000 is borrowed for one year at an interest rate of 1% per month. If this same sum of money is borrowed for the same period at an interest rate of 12% per annum, the saving in interest charges would be:

(a) $ 0
(b) 3
(c) 5
(d) 7
(e) 14

Calculator solution:

At i = 1% per month $F = \$1000(1 + 0.01)^{12} = \$1126.83$

At i = 12% per year $F = 1000(1 + 0.12)^{1} = 1120.00$

Saving in interest = $6.83 ●

Interest table solution:

At i = 1% per month $F = \$1000(F/P,1\%,12) = 1000(1.127) = \$1127.00$

At i = 12% per year $F = 1000(F/P,12\%,1) = 1000(1.120) = 1120.00$

Saving in interest = $7.00 ●

Answer is (d)

ENGR ECON 17

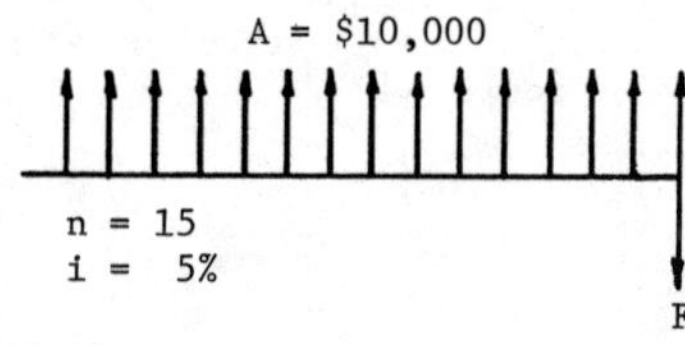

F is closest to

(a) $105,000
(b) 150,000
(c) 157,000
(d) 215,000
(e) 262,000

$F = \$10,000(F/A,5\%,15) = 10,000(21.579) = \$215,790$ ●

Answer is (d)

ENGR ECON 18

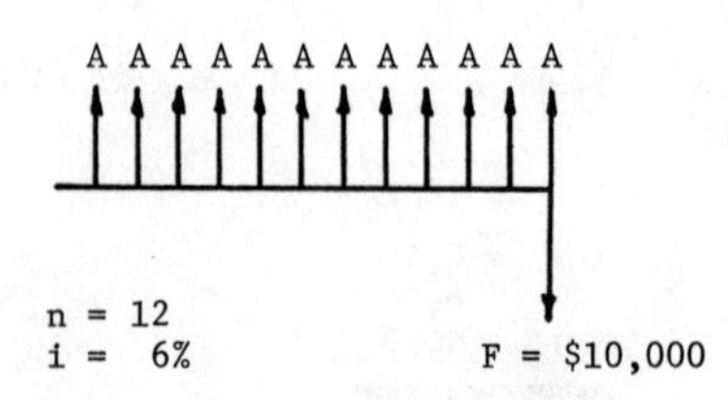

A is closest to

(a) \$ 593
(b) 783
(c) 833
(d) 883
(e) 1193

$A = 10{,}000(A/F,6\%,12) = 10{,}000(0.0593) = \$593$ ●

Answer is (a)

ENGR ECON 19

A company deposits \$1000 every year for ten years in a bank. The company makes no deposits during the subsequent five years. If the bank pays 8% interest, the amount in the account at the end of 15 years is nearest to

(a) \$10,800
(b) 15,000
(c) 16,200
(d) 21,200
(e) 25,200

$F = 1000(F/A,8\%,10)(F/P,8\%,5) = 1000(14.487)(1.469) = \$21{,}281.40$ ●

Answer is (d)

ENGR ECON 20

\$25,000 is deposited in a savings account that pays 5% interest, compounded semiannually. Equal annual withdrawals are to be made from the account, beginning one year from now and continuing forever. The maximum amount of the equal annual withdrawal is closest to:

(a) \$ 625
(b) 1000
(c) 1250
(d) 1265
(e) 1365

Effective interest rate $= (1 + 0.025)^2 - 1 = 0.050625 = 5.0625\%$

Annual withdrawal $A = Pi = 25{,}000(0.050625) = \$1265.63$ ●

Answer is (d)

ENGR ECON 21

What present sum would need to be put in a savings account now to provide a \$1000 annual withdrawal for the next 50 years, if interest is 6%? The present sum is closest to:

(a) \$ 1,000
(b) 10,000
(c) 25,000
(d) 37,500
(e) 50,000

$P = \$1000(P/A,6\%,50) = 1000(15.762) = \$15{,}762$ ●

Answer is (b)

ENGR ECON 22

What present sum is equivalent to a series of $1000 annual end-of-year payments, if a total of 10 payments are made and interest is 6%? The present sum is closest to

(a) $ 6,250
(b) 7,350
(c) 9,400
(d) 10,000
(e) 10,600

$P = 1000(P/A,6\%,10) = 1000(7.360) = \$7360$ ●

Answer is (b)

ENGR ECON 23

A woman made ten annual end-of-year purchases of $1000 of common stock. At the end of the tenth year she sold all the stock for $12,000. What interest rate did she obtain on her investment?

(a) 2%
(b) 4%
(c) 8%
(d) 14%
(e) 20%

$F = A(F/A,i\%,n)$ $\quad 12{,}000 = 1000(F/A,i\%,10)$

$$(F/A,i\%,10) = \frac{12{,}000}{1{,}000} = 12$$

In the 4% interest table
$(F/A,4\%,10) = 12.006$ $\quad$ i is very close to 4% ●

Answer is (b)

ENGR ECON 24

What present sum would be needed to provide for annual end-of-year payments of $15 each, forever? Assume interest is 8%

(a) $120.00
(b) 137.50
(c) 150.00
(d) 187.50
(e) 375.00

In the special situation where n equals infinity, the value of the capital recovery factor is i.

$$A = P(A/P,i\%,\infty) = Pi$$

$$P = \frac{A}{i} = \frac{\$15}{0.08} = \$187.50$$ ●

Answer is (d)

ENGR ECON 25

What amount of money deposited 50 years ago at 8% interest would now provide a perpetual payment of $10,000 per year? The amount is nearest to:

(a) $ 3,000
(b) 8,000
(c) 50,000
(d) 82,000
(e) 125,000

The amount of money needed now to begin the perpetual payments:

$$P' = \frac{A}{i} = \frac{10,000}{0.08} = \$125,000$$

The amount of money that would need to have been deposited 50 years ago at 8% interest is:

$$P = \$125,000(P/F,8\%,50) = 125,000(0.0213) = \$2662.50$$

Answer is (a)

ENGR ECON 26

A dam was constructed for $200,000. The annual maintenance cost is $5000. If interest is 5%, the capitalized cost of the dam, including maintenance, is:

(a) $100,000
(b) 200,000
(c) 215,000
(d) 250,000
(e) 300,000

Capitalized cost is defined as the present worth of perpetual service, and is frequently used in connection with public works projects.

$$\text{Capitalized cost} = \$200,000 + \frac{A}{i} = \$200,000 + \frac{\$5000}{0.05} = \$300,000$$

Answer is (e)

ENGR ECON 27

Given a sum of money Q that will be received six years from now. At 5% annual interest the present worth now of Q is $60. At this same interest rate, what would be the value of Q ten years from now?

(a) $ 60.00
(b) 76.58
(c) 90.00
(d) 97.74
(e) 120.00

This problem illustrates the concept of equivalence. The present sum P = $60 is equivalent to Q six years hence at 5% annual interest.
The future sum F may be calculated by either of two methods:

$$F = Q(F/P,5\%,4) \qquad \text{Eqn (1)}$$

$$\text{or} \quad F = P(F/P,5\%,10) \qquad \text{Eqn (2)}$$

Since P is known, Eqn 2 may be solved directly.

$$F = P(F/P,5\%,10) = \$60(1.629) = \$97.74$$

Answer is (d)

ENGR ECON 28

A piece of property is purchased for $10,000 and yields a $1000 yearly profit. If the property is sold after five years, what is the minimum price to break-even, with interest at 6%?

(a) $ 5,000
(b) 6,554
(c) 7,743
(d) 8,314
(e) 10,000

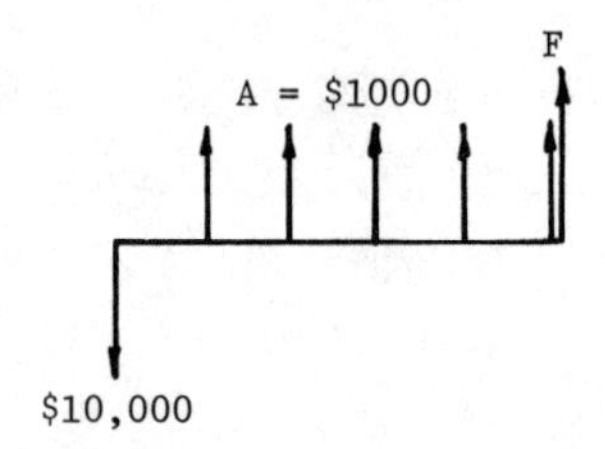

F = $10,000(F/P,6%,5) - $1000(F/A,6%,5)
 = $10,000(1.338) - $1000(5.637) = 13,380 - 5637 = $7743 ●

Answer is (c)

ENGR ECON 29

A steam boiler is purchased on the basis of guaranteed performance. A test indicates that the operating cost will be $300 more per year than the manufacturer guaranteed. If the expected life of the boiler is 20 years and money is worth 8%, how much should the purchaser deduct from the purchase price to compensate for the extra operating cost?

(a) $2945
(b) 3320
(c) 4102
(d) 5520
(e) 6000

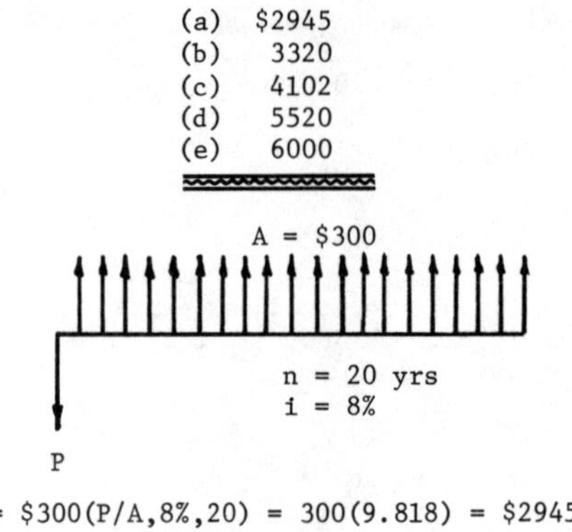

P = $300(P/A,8%,20) = 300(9.818) = $2945.40 ●

Answer is (a)

ENGR ECON 30

Annual maintenance costs for a particular section of highway pavement are $2000. The placement of a new surface would reduce the annual maintenance cost to $500 per year for the first five years and to $1000 per year for the next five years. The annual maintenance after ten years would again be $2000. If maintenance costs are the only saving, what maximum investment can be justified for the new surface? Assume interest at 4%.

(a) $ 5,500
(b) 7,170
(c) 10,000
(d) 10,340
(e) 12,500

Benefits = $1500 per year for the first five years and
$1000 per year for the subsequent five years.
There are several ways of computing the present worth of benefits. Two solutions will be presented.

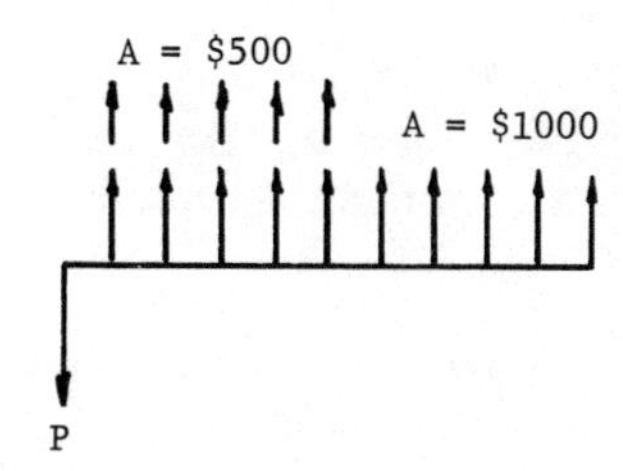

As the sketch indicates, the benefits may be considered as $1000 per year for ten years, plus an additional $500 benefit in each of the first five years.

Maximum investment = Present Worth of benefits
= $1000(P/A,4%,10) + $500(P/A,4%,5)
= $1000(8.111) + $500(4.452)
= $10,337 ●

Alternate Solution.

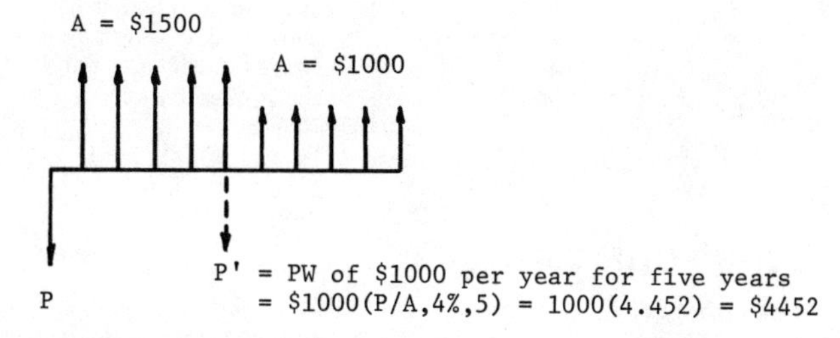

P' = PW of $1000 per year for five years
= $1000(P/A,4%,5) = 1000(4.452) = $4452

Maximum investment = Present Worth of Benefits
= $1500(P/A,4%,5) + P'(P/F,4%,5)
= $1500(4.452) + $4452(0.8219)
= $10,337 ●

Answer is (d)

ENGR ECON 31

A man buys a small garden tractor. There will be no maintenance cost the first year as the tractor is sold with one year's free maintenance. The second year the maintenance is estimated at $20. In subsequent years the maintenance cost will increase $20 per year (that is, 3rd year maintenance will be $40; 4th year maintenance will be $60, and so forth). How much would need to be set aside now at 5% interest to pay the maintenance costs on the tractor for the first six years of ownership?

(a) $101.52
(b) 164.74
(c) 239.36
(d) 284.13
(e) 300.00

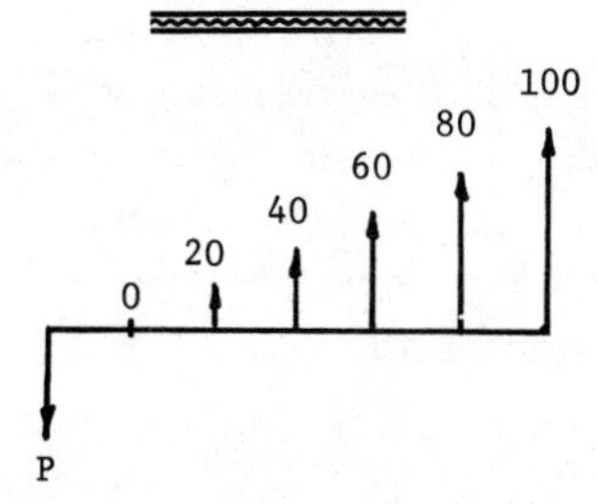

Solution using single payment present worth factors

$$P = 20(P/F,5\%,2) + 40(P/F,5\%,3) + 60(P/F,5\%,4) + 80(P/F,5\%,5) + 100(P/F,5\%,6)$$
$$= 20(0.9070) + 40(0.8638) + 60(0.8227) + 80(0.7835) + 100(0.7462)$$
$$= \$239.35$$ ●

Alternate solution using the gradient present worth factor

$$P = 20(P/G,5\%,6) = 20(11.968) = \$239.36$$ ●

Answer is (c)

ENGR ECON 32

Which one of the following is *NOT* a method of depreciating plant equipment for accounting and engineering economic analysis purposes?

(a) Double entry method
(b) Fixed percentage method
(c) Sum-of-years digits method
(d) Straight line method
(e) Sinking fund method

Answer is (a) ●

ENGR ECON 33

A manufacturing company buys a machine for $50,000. It estimates the machine's useful life is 20 years and that it can then be sold for $5000. Using straight line depreciation, what is the annual depreciation charge?

(a) $2000
(b) 2250
(c) 2500
(d) 2750
(e) 3000

Straight line depreciation

$$\text{Annual depreciation charge} = \frac{P - S}{n} = \frac{50{,}000 - 5{,}000}{20} = \$2250$$ ●

Answer is (b)

ENGR ECON 34

In determining the average annual cost of a proposed project, the formula

$\frac{P - S}{n} + (P - S)(\frac{i}{2})(\frac{n + 1}{n}) + Si$ represents the economic method of

(a) Sinking fund depreciation plus interest on first cost.
(b) Straight line depreciation plus interest on first cost.
(c) Straight line depreciation plus average interest.
(d) Capital recovery with a return.
(e) Amortization plus interest on first cost.

$$\text{Straight line depreciation} = \frac{P - S}{n}$$

$$\text{Average interest} = \frac{\text{First year's interest + Last year's interest}}{2}$$

$$\text{First year's interest} = (P - S)i + Si$$

$$\text{Last year's interest} = \frac{(P - S)}{n}(i) + Si$$

$$\text{Average interest} = \frac{(P - S)i + Si + \frac{(P - S)}{n}(i) + Si}{2}$$

$$= (P - S)\left(\frac{i}{2}\right)\left(\frac{n + 1}{n}\right) + Si$$

Therefore, straight line depreciation plus average interest equals

$$\frac{P - S}{n} + (P - S)\left(\frac{i}{2}\right)\left(\frac{n + 1}{n}\right) + Si \quad ●$$

Answer is (c)

ENGR ECON 35

Company A has fixed expenses of $15,000 per year and each unit of product has a $0.002 variable cost. Company B has fixed expenses of $5000 per year and can produce the same unit of product at a $0.05 variable cost. At what number of units of annual production will Company A have the same overall cost as Company B? Quantity is nearest to

(a) 100,000 units
(b) 200,000
(c) 300,000
(d) 400,000
(e) 2,000,000

Let x = annual production (units)

Total cost to Company A = Total cost to Company B

$$15{,}000 + 0.002x = 5000 + 0.05x$$

$$x = \frac{10{,}000}{0.048} = 208{,}333 \text{ units} \quad ●$$

Answer is (b)

ENGR ECON 36

Plan *A* requires a $100,000 investment now.
Plan *B* requires an $80,000 investment now and an additional $40,000 investment later. At 8% interest, what is the breakeven point on the timing of the additional $40,000 later?

(a) 3 years
(b) 5
(c) 7
(d) 9
(e) 11

The difference between the alternatives is that Plan *A* requires \$20,000 extra now and Plan *B* requires \$40,000 extra n years hence.

At breakeven

$$\$20{,}000 = \$40{,}000(P/F,8\%,n)$$

$$(P/F,8\%,n) = 0.5$$

From the 8% compound interest table (P/F,8%,9) = 0.5002
Therefore, n = 9 years. ●

Answer is (d)

ENGR ECON 37 *(This is a 10-question problem)*

A company is considering buying a new piece of machinery. Two models are available.

	Machine I	*Machine II*
Initial cost	\$80,000	\$100,000
End of useful life salvage value, S	20,000	25,000
Annual operating cost	18,000	15,000 first 10 years; 20,000 thereafter
Useful life	20 yrs	25 yrs

Answer the following ten questions, based on a 10% interest rate.

QUESTION 1 What is the equivalent uniform annual cost for Machine I? The annual cost is closest to

(a) \$21,000
(b) 23,000
(c) 25,000
(d) 27,000
(e) 29,000

$$\begin{aligned} EUAC &= (P - S)(A/P,i\%,n) + Si + \text{Annual operating cost} \\ &= (80{,}000 - 20{,}000)(A/P,10\%,20) + 20{,}000(0.10) + 18{,}000 \\ &= 60{,}000(0.1175) + 2000 + 18{,}000 \\ &= 27{,}050 \end{aligned}$$
●

Answer is (d)

QUESTION 2 What is the equivalent uniform annual cost for Machine II? The annual cost is closest to

(a) \$21,000
(b) 23,000
(c) 25,000
(d) 27,000
(e) 29,000

$$\begin{aligned} EUAC &= (100{,}000 - 25{,}000)(A/P,10\%,25) + 25{,}000(0.10) \\ &\quad + 20{,}000 - 5000(P/A,10\%,10)(A/P,10\%,25) \\ &= 75{,}000(0.1102) + 2500 + 20{,}000 - 5000(6.145)(0.1102) \\ &= \$27{,}379 \end{aligned}$$
●

Answer is (d)

QUESTION 3 The capitalized cost based on Machine I is closest to

(a) $ 60,000
(b) 80,000
(c) 230,000
(d) 270,000
(e) 420,000

Capitalized cost = Present worth of an infinite life.
In Question 1 we computed the Equivalent Uniform Annual Cost (EUAC).

$$P = \frac{A}{i}$$

$$\text{Capitalized Cost} = \frac{A}{i} = \frac{\text{EUAC}}{i} = \frac{\$27{,}050}{0.10} = \$270{,}500$$ ●

Answer is (d)

QUESTION 4 If Machine I is purchased and a fund is set up to replace Machine I at the end of 20 years, the uniform annual deposit that should be made to the fund is nearest to:

(a) $1000
(b) 2000
(c) 3000
(d) 4000
(e) 5000

Required future sum F = $80,000 - 20,000 = $60,000

Annual deposit A = $60,000(A/F,10%,20) = $60,000(0.0175) = $1050 ●

Answer is (a)

QUESTION 5 Machine I will produce an annual savings in material of $25,700 a year. What is the before-tax rate of return if Machine I is installed? The rate of return is closest to:

(a) 6%
(b) 8%
(c) 10%
(d) 20%
(e) 35%

The cash flow for this situation is:

Year	Cash Flow
0	-$80,000
1-20	+25,700 -18,000
20	+20,000

Write one equation with i as the only unknown.

80,000 = (25,700 - 18,000)(P/A,i%,20) + 20,000(P/F,i%,20)

Try i = 8%

80,000 = 7700(9.818) + 20,000(0.2145) = 79,888

Therefore, the rate of return is very close to 8% ●

Answer is (b)

QUESTION 6 Assuming Sum-Of-Years Digits depreciation, what would be the book value of Machine I after two years? Book value is closest to:

(a) $21,000
(b) 42,000
(c) 59,000
(d) 69,000
(e) 79,000

Sum-Of-Years Digits depreciation:

$$\text{Depreciation charge in any year} = \frac{\text{Remaining useful life at beginning of year}}{\text{Sum-Of-Years Digits for total useful life}}(P - S)$$

$$\text{Sum-Of-Years Digits} = \frac{n}{2}(n + 1) = \frac{20}{2}(21) = 210$$

$$\text{1st year depreciation} = \frac{20}{210}(80{,}000 - 20{,}000) = \$\ 5{,}714$$

$$\text{2nd year depreciation} = \frac{19}{210}(80{,}000 - 20{,}000) = \underline{5{,}429}$$

$$\text{TOTAL} = \$11{,}143$$

Book Value = Cost - Depreciation to date
= $80,000 - 11,143
= $68,857 ●

Answer is (d)

QUESTION 7 Assuming Double Declining Balance depreciation, what would be the book value of Machine II after 3 years?

(a) $16,000
(b) 22,000
(c) 58,000
(d) 78,000
(e) 83,000

Double Declining Balance depreciation:

$$\text{Depreciation charge in any year} = \frac{2}{n}(P - \text{Depreciation charges to date})$$

$$\text{1st year depreciation} = \frac{2}{25}(100{,}000 - 0) = \$8{,}000$$

$$\text{2nd year depreciation} = \frac{2}{25}(100{,}000 - 8{,}000) = 7{,}360$$

$$\text{3rd year depreciation} = \frac{2}{25}(100{,}000 - 15{,}360) = \underline{6{,}771}$$

$$\text{TOTAL: } \$22{,}131$$

Book Value = Cost - Depreciation to date
= $10,000 - 22,131
= $77,869 ●

Answer is (d)

QUESTION 8 A new building is being considered to house some equipment. The new building will reduce maintenance costs by \$6000 per year for the first ten years, and \$3000 per year thereafter. Based on a 50-year analysis period, what building construction cost can be justified? The justified construction cost is closest to:

(a) \$ 30,000
(b) 50,000
(c) 90,000
(d) 140,000
(e) 180,000

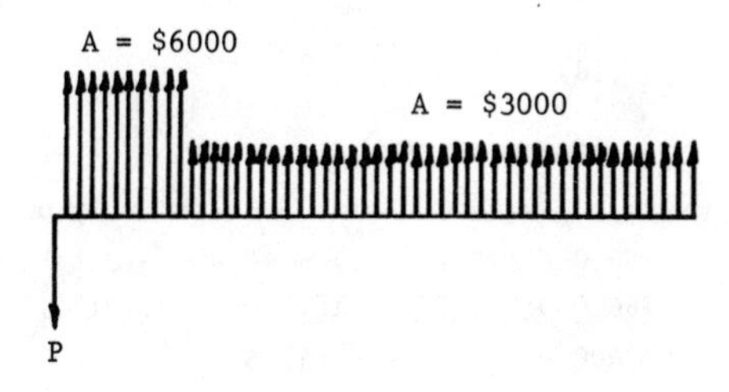

The justified construction cost = Present Worth of savings

$$P = \$6000(P/A,10\%,10) + \$3000(P/A,10\%,40)(P/F,10\%,10)$$
$$= \$6000(6.145) + \$3000(9.779)(0.3855)$$
$$= \$48,179$$ ●

Alternate solution:

$$P = \$3000(P/A,10\%,50) + \$3000(P/A,10\%,10)$$
$$= \$3000(9.915) + \$3000(6.145)$$
$$= \$48,180$$ ●

Answer is (b)

QUESTION 9 The manufacturers of Machine II have announced a price reduction for the machine. What is the breakeven initial price for Machine II, when compared to Machine I?

(a) \$ 91,000
(b) 94,000
(c) 97,000
(d) 100,000
(e) 103,000

In *Question 1* the equivalent uniform annual cost for Machine I was computed to be \$27,050.
For breakeven set the EUAC of Machine I equal to the EUAC of Machine II, and compute initial cost of Machine II at this point.

$$EUAC_I = EUAC_{II}$$

$$\$27,050 = (P - 25,000)(A/P,10\%,25) + 25,000(0.10) + 20,000 - 5000(P/A,10\%,10)(A/P,10\%,25)$$

$$= (P - 25,000)(0.1102) + 22,500 - 5000(6.145)(0.1102)$$

$$= 0.1102P - 2755 + 22,500 - 3386$$

$$\$27,050 = 0.1102P + 16,359$$

$$P = \frac{\$27,050 - 16,359}{0.1102} = \$97,015$$ ●

Answer is (c)

QUESTION 10 What must be the salvage value of Machine I at the end of 20 years for the machine to have an equivalent uniform annual cost of $26,500?

(a) $10,000
(b) 20,000
(c) 30,000
(d) 40,000
(e) 50,000

We can write one equation with S as the only unknown.

$$\$26,500 = \$80,000(A/P,10\%,20) + 18,000 - S(A/F,10\%,20)$$
$$= \$80,000(0.1175) + 18,000 - S(0.0175)$$
$$= \$9400 + 18,000 - 0.0175S$$

$$S = \frac{\$27,400 - \$26,500}{0.0175} = \$51,429$$ ●

Answer is (e)

10

Chemistry

CHEMISTRY 1

All of the following are oxidation-reductions reactions, EXCEPT:

(a) $CaCO_3 \rightarrow CaO + CO_2$

(b) $CO_2 + C \rightarrow 2CO$

(c) $Fe + S \rightarrow FeS$

(d) $2SO_2 + O_2 \rightarrow 2SO_3$

(e) $Hg + I_2 \rightarrow HgI_2$

Oxidation-reduction involves changes in oxidation state of the elements.

In (a): Ca remains at +2, C remains at +4, and O remains at -2. ●

(b): C is oxidized from 0 to +2 and is reduced from +4 to +2 while O remains at -2.

(c): Fe is oxidized from 0 to +2, while S is reduced from 0 to -2.

(d): S is oxidized from +4 to +6, while some of the O is reduced from 0 to -2.

(e): Hg is oxidized from 0 to +2, while I is reduced from 0 to -1.

Answer is (a)

CHEMISTRY 2

To determine the chemical formula of a compound, we need to know all of the following EXCEPT the

(a) elements that compose the compound.
(b) atomic weights of the elements in the compound.
(c) molecular weight of the compound.
(d) percentage composition of the compound.
(e) density of the compound.

All except density are required.

Example: A compound of nitrogen (at. wt. 14) and oxygen (at. wt. 16), of molecular wt. 92, contains 30.5 wt.% nitrogen. Determine its chemical formula.

Basis of calculation: 100 lb of compound

lb moles nitrogen = $\frac{30.5}{14}$ = 2.18 lb moles oxygen = $\frac{69.5}{16}$ = 4.35

Simplest ratio: NO_2 has molecular wt. of 14 + 32 = 46

next higher ratio: N_2O_4 gives correct molecular wt. of 92.

Although the simplest formula is NO_2, the compound exists as the dimer, N_2O_4, nitrogen tetroxide. Other examples of multiples of the simplest formula are P_4 and Hg_2Cl_2.

We note that vapor density may be used to determine molecular weight of gases, while solid density is not directly applicable.

Answer is (e)

CHEMISTRY 3

The number of molecules in 22.4 liters (under standard conditions) of a substance in its gaseous state, is called

(a) Dulong's number
(b) Petit's number
(c) Avogadro's number
(d) Gay-Lussac's number
(e) Graham's number

One mole of any gas under standard conditions occupies 22.4 liters. Each mole of gas has a stated number of molecules. This value, 6.02×10^{23}, is known as the Avogadro number.

Answer is (c)

CHEMISTRY 4

Which of the following metals is the best heat conductor?

(a) aluminum
(b) copper
(c) gold
(d) silver
(e) 1020 steel

In metals there is a relationship between thermal and electrical conductivity. Resistivity is the reciprocal of conductivity.

	Thermal Conductivity, k BTU/hr ft^2 (°F/in)	Electrical Resistivity, ρ ohms(cir mil)/ft
Silver	2900	9.8
Copper	2700	10.3
Gold	2060	14.7
Aluminum	1530	18.5
1020 steel	360	60.

(Gold is used in microelectronics because it does not tarnish.)

Answer is (d)

CHEMISTRY 5

One equivalent weight of H_2SO_4 is equal to

(a) 98.06 g
(b) 2 g
(c) 49.03 g
(d) 96.06 g
(e) 32.06 g

Atomic wts: H = 1.00
S = 32.06
O = 16.00

One equivalent weight of $H_2SO_4 = \frac{2(1) + 32.06 + 4(16)}{2} = 49.03$ g ●

Answer is (c)

CHEMISTRY 6

When exposed to the atmosphere, which of the following liquids is coldest?

(a) oxygen
(b) argon
(c) nitrogen
(d) hydrogen
(e) helium

Liquid	Atmospheric Boiling Point
oxygen	-183°C
argon	-189°
nitrogen	-196°
hydrogen	-253°
helium	-269° ●

Answer is (e)

CHEMISTRY 7

Hard water is water which contains soluble salts of which of the following elements:

(a) Sodium
(b) Sulfur
(c) Calcium
(d) Chlorine
(e) Nitrogen

Soluble salts of calcium and magnesium are generally associated with hardness. The degree of hardness in water, no matter what the cause, is usually expressed in terms of parts per million of calcium carbonate. ●

Answer is (c)

CHEMISTRY 8

Which of the following metals has the highest specific heat capacity at 100°C?

(a) aluminum
(b) bismuth
(c) copper
(d) iron
(e) silver

Dulong & Petit Law

$$\text{atomic } C_p \text{ of metals} \approx \frac{6.4 \text{ BTU/°F}}{\text{lb. mole}}$$

Thus the metal with the lowest atomic weight will have the highest specific heat capacity. Of the group of metals listed, aluminum has the lowest atomic weight. It therefore has the highest specific heat capacity. ●

Answer is (a)

CHEMISTRY 9

Which of the following is a member of the halogen family?

(a) sodium
(b) fluorine
(c) hydrogen chloride
(d) phosphorus
(e) hydrogen

The halogen family ("salt producing") consists of fluorine, chlorine, bromine, and iodine. ●

Answer is (b)

CHEMISTRY 10

A sample of a substance, to which a chemical formula can be assigned, whose weight is equal to its formula weight is termed a:

(a) molecule
(b) mole
(c) gram equivalent weight
(d) one normal solution
(e) atom

Answer is (b) ●

CHEMISTRY 11

The radical ion $(NO_2)^-$ is called a

(a) nitrate
(b) nitrite
(c) nitride
(d) binitrite
(e) binitrate

Answer is (b) ●

CHEMISTRY 12

Hydrogen is common to all

(a) acids
(b) salts
(c) oxides
(d) metals
(e) inorganic material

Answer is (a) ●

CHEMISTRY 13

Which of the following does not illustrate the effect of temperature or pressure on gas solubility?

(a) Air bubbles form on the sides of a warm glass of water.
(b) Soda pop is bottled under pressure.
(c) Boiling frees water of gases.
(d) Air is more humid on rainy days.
(e) Relative humidity is dependent on temperature.

Answer is (d) ●

CHEMISTRY 14

The chemical process which occurs when water is added to cement is

(a) oxidation
(b) Brownian movement
(c) plastic flow
(d) hydration
(e) counter diffusion

Answer is (d) ●

CHEMISTRY 15

An amphoteric hydroxide is one which:

(a) has a valence of -2.
(b) reacts violently with water.
(c) has been created by bombardment with high energy protons.
(d) may act as either a base or an acid in chemical reactions.
(e) decays spontaneously.

Answer is (d) ●

CHEMISTRY 16

All statements about the equilibrium reactions below are correct, EXCEPT:

(a) $N_2 + 3H_2 \rightleftharpoons 2NH_3$ is shifted to the right by pressure increase.

(b) $CaCO_3 \rightleftharpoons CaO + CO_2$ is shifted to the left by pressure increase.

(c) $Ca^{++} + SO_4^{--} \rightleftharpoons CaSO_4$ is shifted to the right by precipitating $CaSO_4$.

(d) $Ag^+ + 2NH_3 \rightleftharpoons Ag(NH_3)_2^+$ is shifted to the right by raising concentration of Ag^+.

(e) $N_2O_4 \rightleftharpoons 2NO_2$ is shifted to the right by pressure increase.

LeChatelier's principle: Systems respond to reduce applied stress. This principle is used to qualitatively estimate the effect of altered solution concentrations or pressures on chemical equilibria. Increasing reactant concentration produces more product (to reduce reactant concentration); raising pressure favors fewer gas molecules (to reduce pressure); and vice versa.

The false statement is (e) ●

CHEMISTRY 17

All of the following statements about pH of a solution are correct, EXCEPT:

(a) It is $-\log_{10}[H^+]$

(b) It is held reasonably constant in buffer solutions.

(c) pH 10 is ten times more alkaline than pH 9.

(d) pH of tap water is below 7 due to dissolved CO_2.

(e) It is zero for neutral solutions.

---

pH is $-\log_{10}[H^+]$ Hydrogen ion concentration is gram moles/liter over the range from 1 to 10^{-14} is expressed by pH from 0 to 14. Buffer solutions contain salts of weak acids or weak bases; they stabilize pH of solutions despite dilution or small additions of acid or base. pH of acid solutions is below 7; of neutral solutions is equal to 7; and of alkaline solutions is above 7.

The false statement is (e) ●

11

Structure of Matter

STRUCTURE OF MATTER 1

The element tin has ten different stable isotopes. The atomic nuclei of all isotopes have the same:

(a) number of neutrons.
(b) number of protons.
(c) radius.
(d) mass.
(e) binding energy.

Nuclei of the various isotopes of tin have the same atomic number Z = 50 protons but different mass numbers, A, ranging from 112 to 124 because of differing A-Z number of neutrons. Nuclear radii and configurations differ. Binding energy is derived from the mass discrepancy between mass of the whole nucleus and summation of masses of its individual nucleons (protons and neutrons). One atomic mass unit (amu) equals 931 Mev by the $E = mc^2$ energy-mass equivalence, and one amu is 1/12 the mass of an atom of ${}_6C^{12}$ (the basis of atomic weights) or 1.66×10^{-24} gram. Binding energy per nucleon varies with mass number, averaging 8 Mev/nucleon, but being much less for the lightest, and slightly less for the heaviest nuclei. Energy release and stability gained through thermonuclear fusion of light elements and fission of the heaviest elements are consequences of binding energy differences.

Answer is (b) ●

STRUCTURE OF MATTER 2

All of the following statements about the electron structure of atoms are correct, EXCEPT:

(a) Electrons of highest energy are located in the outer orbitals.
(b) Each orbital can contain one electron.
(c) K, L, M, etc. shells are designations representing the principal quantum numbers n = 1, 2, 3 etc.
(d) Each shell may contain up to n^2 orbitals, but not exceeding 16.
(e) One s, three p, five d and seven f orbitals, containing up to 2, 6, 10 and 14 electrons respectively, have different spatial configurations.

Four types of quantum numbers describe the building up of shells of electron orbitals with discretely increasing energy level for each orbital.

Principal quantum number, n, represents a major grouping of energy levels from 1 to 7, sometimes referred to as K, L, M etc. shells.

Azimuthal quantum number, ℓ, denotes different shapes of orbitals within the major groupings. These shapes are labeled s, p, d and f for $\ell = 0, 1, 2$ and 3 respectively.

Magnetic quantum number, m, may be +1, -1 or 0. Each value of m defines a magnetic property of the electron resulting from its orbital motion.

Spin magnetic quantum number, s, may have a value of +1/2 or -1/2 depending on direction of electron spin. This permits each orbital to contain two electrons of opposite spin.

One s orbital, three p orbitals, five d orbitals and seven f orbitals exist. Each orbital can contain two electrons. The maximum number of electrons in each major energy level ($n = 1, 2, 3, \ldots$) is $2n^2$, but not exceeding 32.

These numbers identify an overlapping succession of electron orbitals designated, in order of increasing energy: 1s, 2s, 2p, 3s, 3p, 4s, 3d, 4p, 5s, 4d, 5p, 6s, 4f, 5d, ...

Note that s orbitals exist at $n = 1$ upward, p orbitals at $n = 2$ upward, d orbitals at $n = 3$ upward, etc.

Energy levels are filled sequentially, beginning with the lowest 1s level. In the ground state the lowest energy levels are filled and no vacancies exist.

Exited states result when an electron receives energy, vacates a lower energy orbital, and occupies a higher energy orbital.

One photon of discrete energy is emitted or absorbed as an electron makes transition between differing energy levels of the various orbitals.

The false statement is (b) ●

STRUCTURE OF MATTER 3

All of the following elementary particles have detectable mass, EXCEPT:

(a) neutron n^o
(b) proton p^+
(c) electron e^-
(d) positron e^+
(e) neutrino ν

The field of high energy particle physics has shown the existence of more than 30 elementary particles, including all of the above plus various mesons, hyperons and other particles. The neutrino is a massless, unchanged particle that can possess both energy and momentum; it is emitted together with an electron during beta decay. The proton p^+ (and antiproton p^-), electron e^- (and positron e^+) and neutrino ν (and antineutrino $\bar{\nu}$) are stable outside the nucleus and do not decay prior to reaction; the neutron n^o has a half life of 12 minutes. All the others have lifetimes much less than 10^{-6} second and decay spontaneously into other particles, react with other particles, or are annihilated with energy release on combination with their antiparticles.

Answer is (e) ●

STRUCTURE OF MATTER 4

Which of the following modes of radioactive decay results in an increased Z or number of protons in the nucleus?

- (a) α decay (helium nucleus emission)
- (b) negative beta decay (electron emission)
- (c) positive beta decay (positron emission)
- (d) K electron capture (unchanged neutrino emission)
- (e) γ emission (photon energy emission)

Radioactive decay is summarized below:

Radioactive decay mode	net N = A-Z, change in no. of neutrons	net Z, change in no. of protons	Reaction within nucleus
α decay	-2	-2	emission of $({}_2He^4)^{++}$ reduces nucleus by 2 neutrons and 2 protons
negative beta decay	-1	+1	$n^o \rightarrow p^+ + e^- \nearrow + \bar{\nu} \nearrow$ electron and antineutrino emitted
positive beta decay	+1	-1	$p^+ \rightarrow n^o + e^+ \nearrow + \nu \nearrow$ positron and neutrino emitted
electron capture	+1	-1	$p^+ + e^- \rightarrow n^o + \nu \nearrow$ neutrino emitted
γ emission	0	0	excited nucleus loses excess energy by energetic photon emission

Radioactive nuclei spontaneously decay and transform themselves into other nuclear species with a more stable neutron-proton ratio and configuration. Often a succession of α- and β-decays, with accompanying γ-emission, are required before a nucleus reaches stability. In electron capture the nucleus captures an inner K orbital electron and emits a neutrino, accompanied by x rays due to electronic shifts characteristic of the new atom. Electron capture and positive beta decay (positron emission) yield the same product, though positron emission is rare and does not occur in the naturally radioactive isotopes.

If ${}_{53}I^{131}$ could hypothetically decay by each of the listed modes the daughter product formed and radiation emitted would be:

α decay: ${}_{51}Sb^{127} + {}_2He^4$

negative beta: ${}_{54}Xe^{131} + e^- + \bar{\nu}$

positive beta: $_{52}Te^{131} + e^{+} + \nu$

electron capture: $_{52}Te^{131} + \nu$

γ emission: $_{53}I^{131} + \gamma$

(The actual decay mode is mostly by negative beta decay to stable $_{54}Xe^{131}$.)

Answer is (b) ●

STRUCTURE OF MATTER 5

$_{38}Sr^{90}$ decays by negative beta emission and has a halflife of 28 years. All of following statements are correct, EXCEPT:

(a) Mean lifetime of $_{38}Sr^{90}$ is 40 years.

(b) Activity of 10 micrograms of $_{38}Sr^{90}$ is 1.4 millicuries.

(c) First daughter product of the beta decay is $_{39}Y^{90}$.

(d) Decay probability per year for any atom of $_{38}Sr^{90}$ is numerically equal to the decay constant of 0.025/year.

(e) After 100 years more than 10% of the original $_{38}Sr^{90}$ remains.

Radioactive decay is statistical in nature and follows an exponential law.

Fraction remaining is equal to $e^{-\lambda t}$ where λ is the decay constant.

Halflife $t_{\frac{1}{2}}$ is related to decay constant by $\lambda = \frac{\ln 2}{t_{\frac{1}{2}}} = \frac{0.693}{t_{\frac{1}{2}}}$

$$\lambda = \frac{0.693}{28} = 0.025/\text{year}.$$

Mean lifetime of all of the atoms, $t = \frac{1}{\lambda} = 1.44 t_{\frac{1}{2}}$,

thus mean lifetime of $_{38}Sr^{90}$ is 1.44(28) = 40. years.

Activity, number of disintegrations/sec, is expressed in curies.

One curie equals 3.70×10^{10} disintegrations/sec.

Activity = λN where λ is decay constant/sec and N is number of atoms of isotope present.

$$\lambda = \frac{0.025}{\text{year}} \times \frac{\text{year}}{365 \text{ days}} \times \frac{\text{day}}{24 \text{ hrs}} \times \frac{\text{hour}}{3600 \text{ sec}} = 7.9 \times 10^{-10}/\text{sec}.$$

$$N = \frac{10 \times 10^{-6} \text{ grams}}{90 \text{ grams/gm atom}} \times 6.025 \times 10^{23} \text{ atoms/gm atom} = 6.7 \times 10^{16} \text{ atoms present}$$

$$\text{Activity} = \lambda N = (7.9 \times 10^{-10}/\text{sec})(6.7 \times 10^{16} \text{ atoms}) = 5.3 \times 10^{7} \text{ disintegrations per second}$$

$$= \frac{5.3 \times 10^{7}}{3.70 \times 10^{10}} = 1.4 \times 10^{-3} \text{ curies or 1.4 millicuries.}$$

Thus, short halflives yield more activity, disintegrations/sec.

Since negative beta decay converts one neutron to a proton and ejects an electron, the atomic number increases by one to 39, the name changes accordingly from 38 = Strontium to 39 = Yttrium, and the mass number, or total neutrons plus protons, remains unchanged at 90.

Result: first daughter product is ${}_{39}Y^{90}$.

Decay constant is almost exactly the probability that any one atom will decay in the time interval. One half of the original amount remains after one halflife, $\frac{1}{4}$ after two halflives, $\frac{1}{8}$ after three halflives, and a trivial amount remains forever. After 100 years, the fraction remaining is $e^{-.025(100)} = e^{-2.5} = 0.082$

The false statement is (e) ●

STRUCTURE OF MATTER 6

Radioactive ${}_{6}C^{14}$ decays at a rate proportional to the amount present at any time. Its halflife is 5600 years. If the amount of ${}_{6}C^{14}$ present in wood charcoal taken from a ruin is 40% of the amount in timber growing now, estimated age of the charcoal is nearest to:

(a) 4200 years
(b) 5600 years
(c) 7400 years
(d) 8400 years
(e) 9500 years

Decay constant λ is related to halflife by

$$\lambda = \frac{\ln 2}{t_{\frac{1}{2}}} = \frac{0.693}{t_{\frac{1}{2}}} = \frac{0.693}{5600} = 1.24 \times 10^{-4}/\text{year}.$$

Alternately, the decay constant λ can be stated as a disintegration rate of one part in about 8000 per year.

Fraction present = $e^{-\lambda t}$, where t is time elapsed.

$$0.40 = e^{-1.24 \times 10^{-4} t} \qquad 1.24 \times 10^{-4} t = \text{antiln}_e\ 0.40 = 0.915$$

$$t = \frac{0.915}{1.24 \times 10^{-4}} = 7400 \text{ years}$$ ●

Answer is (c)

STRUCTURE OF MATTER 7

Which of the following nuclear reactions is thermonuclear?

(a) ${}_{7}N^{14} + {}_{o}n^{1} \longrightarrow {}_{6}C^{14} + {}_{1}H^{1}$

(b) ${}_{5}B^{10} + {}_{o}n^{1} \longrightarrow {}_{3}Li^{7} + {}_{2}He^{4}$

(c) ${}_{92}U^{238} + {}_{o}n^{1} \longrightarrow {}_{92}U^{239} + \gamma$

(d) ${}_{1}H^{3} + {}_{1}H^{2} \longrightarrow {}_{2}He^{4} + {}_{o}n^{1}$

(e) ${}_{92}U^{235} + {}_{o}n^{1} \longrightarrow$ 2 fission fragments + 1 to 3 ${}_{o}n^{1}$

Nomenclature for nuclear reactions uses $_Z(\text{name})^A$ for nuclei and particles.

Z is atomic number corresponding to name, and A is mass number.

Thus: proton (p^+) is shown as $_1H^1$, neutron (n^o) as $_0n^1$, and the alpha particle (α) as $_2He^4$.

Superscripts and subscripts must balance for all valid reactions. Although all nuclear reactions involve conservation of mass-energy, electrical charge and number of nucleons, skeleton reactions are usually written as above. Type reactions may be further abbreviated showing the input-output particles or emissions:

- n - p for neutron absorption and proton emission
- n - α for neutron absorption and α emission
- n - γ for neutron adsorption and γ emission
- n - f for neutron adsorption and fission

(a) is an n - p reaction that produces $_6C^{14}$ in the upper atmosphere when neutrons from cosmic ray events bombard atmospheric nitrogen.

(b) is an n - α reaction used both for detection of thermal neutrons and for reactor control.

(c) is an n - γ neutron capture reaction using abundant $_{92}U^{238}$ to produce $_{92}U^{239}$ which decays to $_{94}Pu^{239}$ in a few days.

(d) is the triton - deuteron thermonuclear reaction which can be maintained only at extremely high temperature and pressure. Energy release is 17.6 Mev per event. Similar reactions involving fusion of light nuclei into heavier ones occur in the sun and stars. ●

(e) is an n - f fission reaction that produces two distributions of fission fragments in the ranges of 84 - 104 and 130 - 149 with maxima at 93 and 134. It is accompanied by a competing n - γ reaction that can consume more than half of the average 2.5 neutrons released per event.

Answer is (d)

STRUCTURE OF MATTER 8

Which of the following isotopes would be expected as product of an α - n reaction on $_4Be^9$?

(a) $_5B^9$

(b) $_4Be^{10}$

(c) $_6C^{12}$

(d) $_2He^6$

(e) $_5B^{12}$

(a) results from a p - n reaction, (b) from n - γ, (c) from α - n
(d) from n - α , and (e) from α - p.

$$_4Be^9 + {}_2He^4 \longrightarrow {}_6C^{12} + {}_0n^1$$ ●

Answer is (c)

STRUCTURE OF MATTER 9

All of the following statements about cross section, σ, are correct, EXCEPT:

(a) Cross section is expressed in barns (10^{-24} cm^2).
(b) Cross section indicates the likelihood of interaction between an incident particle and target nucleus.
(c) One barn is of the order of magnitude of the geometrical cross section of the nucleus.
(d) Maximum cross section is about 100 barns.
(e) Cross sections vary with type reaction and energy of incident particles, often having peaks at specific energies.

Cross section σ may be subscripted:

σ_a indicating absorption (capture, or n - γ reaction),

σ_s for scattering,

σ_f for fission,

and for various other reactions, such as

σ_p for n - p reaction.

Competing reactions each have their own cross section which varies widely with incident particle energy. Examples for thermal energy neutrons:

(1) $_{92}U^{235}$ has σ = 683 for n - γ capture, and σ_f = 577 for fission.

(2) $_{48}Cd^{113}$ has σ = 2.1×10^4 and is used for reactor control.

(3) $_8O^{16}$ has σ = 2×10^{-5}, hence is very unreactive with thermal neutrons.

(4) $_1H^2$ deuterium (D) has σ = 5.7×10^{-4}, $_6C^{12}$ has σ = 3.3×10^{-3}

while $_1H^1$ hydrogen has σ = 0.33

Therefore heavy water D_2O is an excellent moderator for slowing fast neutrons to thermal energy in reactors because energy is dissipated on collision and reaction seldom occurs.

The false statement is (d) ●

STRUCTURE OF MATTER 10

All of the following are characteristic of waves, EXCEPT:

(a) Energy of a wave varies with the square of its amplitude.

(b) Interference is interaction of waves. The observed effect is based on the principle of superposition.

(c) Diffraction, or bending of a wave, occurs around the edges of obstacles.

(d) Refraction, or bending of a beam, occurs when it passes between media with different velocities of propagation.

(e) Wave frequency changes on entering a medium where the velocity of propagation is changed.

Energy may be dissipated by absorption (attenuation) during passage through a medium. Conservation of energy requires that waves radiating outward from the source must deliver lesser amounts of energy per unit area since the receiving area increases with distance squared. Energy contained in most waves varies with the square of the amplitude of displacement from equilibrium.

Refraction, reflection, interference, and diffraction are characteristic of all wave motion. Diffraction effects become evident when obstacle dimensions are comparable with wavelength.

Frequency remains unchanged in passing between media. ●

The false statement is (e)

STRUCTURE OF MATTER 11

All of the statements about electromagnetic waves are correct, EXCEPT:

(a) Electromagnetic waves are transverse waves that consist of mutually coupled, changing electric and magnetic fields.

(b) Directions of the electric field, of the magnetic field, and of propagation are mutually perpendicular.

(c) Field intensity changes take place, but no displacement of mass occurs in the path of electromagnetic waves.

(d) Electric field maxima and minima occur out of phase with those of the associated magnetic field.

(e) Velocity of propagation is dependent only on permittivity, ϵ , and permeability, μ , of the medium.

Polarization proves that electromagnetic waves are transverse, for longitudinal waves cannot be polarized. Unlike electromagnetic waves, other types of waves cause displacement of mass. Electromagnetic waves are propagated in vacuum and in dielectric materials, but are attenuated due to I^2R losses in conductors. Electric field maxima and minima occur simultaneously and in phase (except very close to the oscillator).

Velocity of propagation $v = \sqrt{1/\epsilon\mu}$ in materials is less than speed of light in vacuum, $c = 3 \times 10^8$ meters/sec.

The false statement is (d) ●

STRUCTURE OF MATTER 12

All of the following statements about sound are correct, EXCEPT:

(a) Sound is propagated by longitudinal compressional waves.

(b) Velocity of sound in solids may be calculated from their elastic modulus and density.

(c) Velocity of sound in gases is independent of pressure.

(d) Sound waves do not exhibit refraction in passing between media.

(e) Velocity of wave propagation equals the product of frequency and wavelength.

Velocity of longitudinal sound waves in solids may be calculated from elastic modulus and density: v in ft/sec $= \sqrt{\frac{E\, g_c}{\rho\, 12}}$ where E is tensile-conpressional (Youngs) modulus in psi, ρ is density in lb/cu in, with $g_c = 32.2$ and 12 in/ft being inserted for correct dimensions. Shear modulus G replaces E for transverse waves. Bulk modulus of elasticity K replaces E for liquids. In gases $v = \sqrt{k\, g_c\, R\, T}$ where $k = C_p/C_v$, R is gas constant in ft $lb_f/lb_m\,°R$, and T is absolute temperature in °R.

Approximate velocity of sound at ambient temperature is 1130 ft/sec in air, 4720 ft/sec in water and 15,800 ft/sec in steel.

Both longitudinal and electromagnetic waves exhibit reflection, refraction, diffraction, and interference. Sound waves are refracted (bent) in passing between media in a manner analogous to light waves; Snells law applies. Index of refraction is the ratio of velocities in the two media. Frequency remains constant and wavelength varies because $v = f\lambda$. ●

The false statement is (d)

STRUCTURE OF MATTER 13

All of the following statements about sound intensity are correct, EXCEPT:

(a) Sound intensity level is energy expressed in decibels,

$$db = 10 \log_{10}\left(\frac{I}{I_o}\right)$$

(b) Zero level of sound intensity, I_o, is established at an energy of 10^{-16} watt/cm^2.

(c) Two sound sources, each individually of 70 db sound intensity, together produce an intensity of 80 db.

(d) Intensity of the lower audible limit at 50 cycles/sec is approximately 10^6 times (i.e., 60 db above) that existent at 3000 cycles/sec.

(e) The human ear is most sensitive to sound at 3000 cycles/sec, and is responsive in young adults from about 20 to 20,000 cycles/sec.

Sound intensity or actual energy per unit area is expressed on the logarithmic decibel scale with the zero level at 10^{-16} watt/cm^2. One sound source of 70 db has an energy 10^7 times as intense as reference, two identical 70 db sound sources have energy 2×10^7 times as intense as reference, or 73 db above reference. One 70 db source plus one 75 db source have combined energies of

1×10^7 plus 3.16×10^7 or 4.16×10^7 times reference.

Since $\log_{10}$ of 4.16×10^7 is 7.62, the combined energy is 76.2 db above reference.

Loudness of noise, being dependent upon the frequency response and sensitivity of the human ear, also is expressed on scales of loudness level in phons, and of loudness sensation in millisones.

The false statement is (c) ●

STRUCTURE OF MATTER 14

At which frequency in cycles/sec does an observer hear a 1000 cycle/sec tone if both source and observer are each moving towards each other at 88 ft/sec? Sound velocity is 1100 ft/sec in air.

(a) 852
(b) 926
(c) 1080
(d) 1088
(e) 1174

The Doppler effect is an apparent frequency shift experienced by the observer due to relative motions. This applies to both light and sound, and explains the spectral red shift observed astronomically.

A moving source compresses or expands relative wavelength experienced by the observer, and a moving observer increases or decreases relative propagation velocity of the wave train. $v = f\lambda$ applies.

	Relative v in ft/sec observed due to motion of observer	Relative λ in ft observed due to motion of source	$f = v/\lambda$ frequency observed in cycles/sec
Stationary source Moving observer S ←0 S 0→	increased or decreased v 1100 + 88 = 1188 ft/sec 1100 - 88 = 1012 ft/sec	constant λ $\frac{1100}{1000} = 1.100$ 1.100	 $\frac{1188}{1.100} = 1080$ 920
Moving source Stationary observer S→ 0 ←S 0	constant v 1100 1100	compressed or expanded λ 1.100(1012/1100) = 1.012 1.100(1188/1100) = 1.118	 1088 926
Moving source Moving observer S→ ←0 S→ 0→ ←S ←0 ←S 0→	increased or decreased v 1188 1012 1188 1012	compressed or expanded λ 1.012 1.012 1.188 1.188	 1174 ● 1000 1000 852

Answer is (e)

STRUCTURE OF MATTER 15

Frequency f of monochromatic yellow light which has a wavelength λ of 589 nm (nanometers) in vacuum is nearest to:

(a) 295,000 GHz
(b) 510,000 GHz
(c) 589,000 GHz
(d) 1,020,000 GHz
(e) 1,180,000 GHz

$kHz = 10^3$ Hz $= 10^3$ cycles/sec, $MHz = 10^6$ Hz, $GHz = 10^9$ Hz.

nanometer $= 10^{-9}$ meter = millimicron = 10 Å

$= 589$ nm $= 589 \times 10^{-9}$ m $= 5890$ Å, $v = f\lambda = 3.00 \times 10^8$ m/sec

3.00×10^8 m/sec $= f(589 \times 10^{-9}$m)

$$f = \frac{3.00 \times 10^8}{589 \times 10^{-9}} = 5.1 \times 10^{14} \text{ cycles/sec} = 510{,}000 \text{ GHz}$$ ●

Answer is (b)

STRUCTURE OF MATTER 16

A radio station broadcasts at a frequency of 600 kilohertz. The corresponding wave length is closest to which of the following values?
Assume the velocity of sound is 300 meters/sec and the velocity of light is 3×10^8 meters/sec.

(a) 500 meters
(b) 2×10^{-3} meters
(c) 5×10^{-4} meters
(d) 2000 meters
(e) 50 meters

Speed of a wave = Frequency x Wavelength $v = f\lambda$

$$\lambda = \frac{v}{f} = \frac{3 \times 10^8 \text{ meters/sec}}{600{,}000\text{/sec}} = 500 \text{ meters}$$ ●

Answer is (a)

STRUCTURE OF MATTER 17

All of the following statements about photons are correct, EXCEPT:

(a) Photons are small quanta of electromagnetic energy.
(b) Energy of photons varies according to $E = hf$, where h is Planck's constant, and f is frequency.
(c) Photon energy varies with source intensity, i.e. a bright light source emits photons of higher energy than a dim source.
(d) Velocity of photons in vacuum is c, the speed of light, 3×10^{10} cm/sec.
(e) Two light sources of the same wavelength emit photons of the same energy.

The discrete energy per photon is independent of intensity or number of photons emitted.

$E = hf$ where $h = 6.625 \times 10^{-34}$ joule-sec.

Velocity v, frequency f, and wavelength λ are related by $v = f\lambda$.

The false statement is (c) ●

STRUCTURE OF MATTER 18

A ray of light passes from air into water. The angle of incidence is 30 degrees from the vertical. Assume the velocity of light in air and vacuum to be 3×10^{10} cm/sec and in water to be 2.26×10^{10} cm/sec. Find the angle of refraction relative to the vertical.

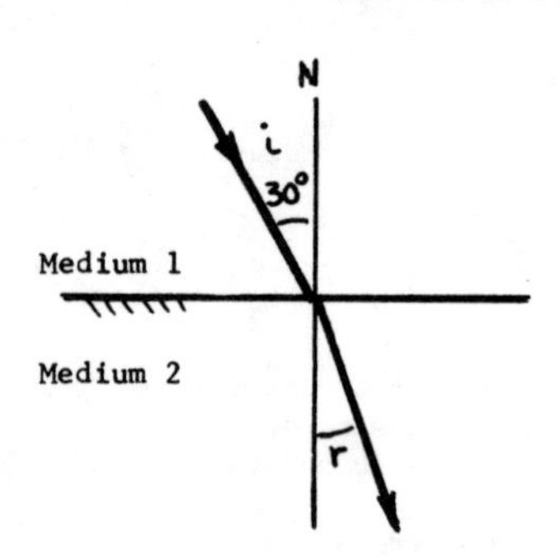

A ray of light undergoes an abrupt change of direction when passing obliquely into another medium in which it travels at a different velocity. The ray will be deviated toward the normal (vertical in the illustration) when the velocity is reduced (the situation in this problem) and deviated away from the normal when the velocity is increased.

The law of refraction:
When a wave travels obliquely from one medium into another, the ratio of the sine of the angle of incidence to the sine of the angle of refraction is the same as the ratio of the respective wave velocities in these media, and is a constant for two particular media.

Thus:

$$\frac{\sin i}{\sin r} = \frac{V_1}{V_2} \qquad \sin r = \frac{V_2 \sin i}{V_1} = \frac{2.26 \times 10^{10} \times 0.5}{3 \times 10^{10}} = 0.3767$$

Angle of Refraction (r) = 22.1 degrees ●

Total internal reflection can occur when an internal ray is reflected at an interface. This occurs when incident angle exceeds a critical angle c, determined from sin c = n of new media/n of old media, and is limited to situations where new media is less dense (has lower n) than old media.
Example: a ray in glass may be reflected rather than penetrating a glass-air interface.

STRUCTURE OF MATTER 19

A one-candlepower light source radiates luminous flux at the rate of

(a) 1 lumen
(b) 2 lumens
(c) π lumens
(d) 2π lumens
(e) 4π lumens

A one-candlepower source radiates 1 lumen upon a surface of 1 square foot at a distance of 1 foot. If one imagines a spherical shell of 1 foot radius around the 1 candlepower source, it is clear that each square foot receives 1 lumen. The area of the spherical shell is $4\pi R^2$, therefore the 1 candlepower source radiates 4π lumens. ●

Answer is (e)

12

Computer Programming

PROGRAMMING 1

Based on the FORTRAN statements, how many input data cards will be read?

```
    READ(60,10)L,(X(I),I=1,13)
 10 FORMAT(I5/(6F13.2))
```

(a) 1
(b) 2
(c) 3
(d) 4
(e) 14

[Please note that throughout this chapter the card reader is assigned logical unit number 60 and the line printer 61. This will vary from computer center to computer center. Often, for example, 5 and 6 are used. Otherwise, standard FORTRAN IV syntax is assumed.]

In this problem one value will be read from the first data card, and six values from subsequent data cards until the read statement is satisfied. It will therefore take four data cards to read the 14 values. ●

Answer is (d)

PROGRAMMING 2

How many data cards will be read by the following

```
    DIMENSION L(43)
    READ(60,15)B,(L(I),I=1,33)
 15 FORMAT(F10.0/(5I5))
```

(a) 1
(b) 2
(c) 7
(d) 8
(e) 34

The value of B will be read from the first data card. Five values of L will be read from each subsequent data card until the read statement is satisfied. ●

Answer is (d)

PROGRAMMING 3

Consider the following

```
   READ(60,12)N
12 FORMAT(I3)
   M=1
   DO 10 I=2,N
10 M=M*I
   WRITE(61,12)M
   END
```

What is the relationship between input N and the output M?

(a) $M = N$

(b) $M = 2N^2$

(c) $M = N^{N-1}$

(d) $M = N!$

(e) $M = N^N$

Answer is (d) ●

PROGRAMMING 4

Consider the following

```
   A=14.5
15 IF(A.GT.10.)GO TO 30
20 GO TO 25
30 GO TO 35
```

The label of the next statement to be executed following this portion of the program is

(a) 15
(b) 20
(c) 25
(d) 30
(e) 35

This is the one branch logical IF statement. If the logical expression within the parentheses is *TRUE*, the statement to the right will be executed; otherwise it will not be executed.

In this problem 14.5 is Greater Than 10, hence the expression is true and one will go to the statement labelled 30 and then to statement 35. ●

Answer is (e)

PROGRAMMING 5

Consider the following

```
ID=3**2/3.3+1
GO TO(10,11,12,13,14),ID
```

The label of the next statement to be executed is

(a) 10
(b) 11
(c) 12
(d) 13
(e) 14

The hierarchy of arithmetic operations is

1. Exponentiation
2. Multiplication and Division
3. Addition and Substraction

The order of operations is from group 1 to group 2 to group 3. Within any group evaluation proceeds from left to right. Parentheses may be used to override this order of operations.

ID = $3^2/3.3 + 1 = 2.73 + 1 = 3.73$ which truncates to 3

The next statement to be executed is 12 ●

Answer is (c)

PROGRAMMING 6

Consider the following

```
J=2*3**2/3.3-1.2*3+1
GO TO(5,4,3,6,2),J
```

What is the label of the next statement to be executed?

(a) 2
(b) 3
(c) 4
(d) 5
(e) 6

Mathematically the computation is

$$[2(3^2)]/3.3 \ -(1.2*3) +1 = 2.85 \text{ which truncates to } 2$$

Answer is (c) ●

PROGRAMMING 7

In the following statements select the one which is not standard FORTRAN.

(a) GO TO(1,2,3,2),N

(b) DO 50,I=1,5

(c) GO TO 15

(d) WRITE(61,20)

(e) C(I)=A(I+1)**B(I)

The comma after 50 in (b) is not standard FORTRAN. The statement is acceptable, however, for use with some FORTRAN compilers.
Item (d) has a blank list. This often occurs when printing output headings, etc.

Answer is (b) ●

PROGRAMMING 8

In the following FORTRAN statements select the one that is incorrectly written.

(a) X = 2,412.62

(b) A = 412

(c) IF(3.*A.EQ.B) GO TO 20

(d) B=A**I

(e) IF(X)10,20,10

A comma must not be used in numbers like 2412.62 ●

Answer is (a)

PROGRAMMING 9

Which one of the following is not a FORTRAN declaration?

(a) REAL
(b) INTEGER
(c) COMMON
(d) EQUIVALENCE
(e) END

The END statement signals the compiler the physical end of either a subprogram or the main program. The other four items are nonexecutable statements (declarations).

Answer is (e) ●

PROGRAMMING 10

(This is a 4-part problem)

Consider the following FORTRAN program.

```
Line
No.
 1          PROGRAM SORT
 2          DIMENSION LIST (100)
 3          READ(60,10) N
 4      10  FORMAT(I3)
 5          READ(60,15) (LIST(I), I=1,N)
 6      15  FORMAT(5X,10I4)
 7      20  DO 25 K=2,N
 8          J=K-1
 9          IF(LIST(K).GE.LIST(J)) GO TO 25
10          NTEMP = LIST(J)
11          LIST(J) = LIST(K)
12          LIST(K) = NTEMP
13      25  CONTINUE
14          N= N-1
15          IF(N.GT.1) GO TO 20
16      30  WRITE(61,15) (LIST(I), I=1,N)
17          END
```

Answer the following four questions assuming that N = 5.

QUESTION 1 The LIST is sorted to

(a) put the even numbers first.
(b) put the odd numbers first.
(c) put the numbers in increasing order.
(d) put the numbers in decreasing order.
(e) make no change in the LIST.

This program is a push-down sort. The sort begins by comparing the two values at the top of the list. If a larger value is above a smaller value, they are reversed in the list. As the sort proceeds, larger values are moved downward (and smaller values moved upward) so the largest value encountered moves to the bottom of the list sorted. In the next iteration the list to be sorted is shortened (N= N-1) and the process repeated. This continues until the list has been shortened to one item - at which time the sort is completed. The sorted list will be ordered with the smallest value at the top and the largest value at the bottom. ●

Answer is (c)

QUESTION 2 How many times is the statement in Line 8 executed?

(a) 4
(b) 10
(c) 14
(d) 15
(e) 20

The DO loop at line 7 causes line 8 to be executed a number of times for each value of K.
With N=5, K=2, 3, 4, 5.

When	*Line 8 is executed*
N = 5	4 times
4	3
3	2
2	1
	10 ●

Answer is (b)

QUESTION 3 The READ statement in Line 5 will cause how many data cards to be read?

(a) 1
(b) 2
(c) 3
(d) 4
(e) 5

Answer is (a) ●

QUESTION 4 Which one of the following is untrue?

(a) The same FORMAT statement may be used for both a READ and a WRITE statement.

(b) Line 10 could not be omitted from the program and the same results be obtained.

(c) If the expression in line 15 were changed from (N.GT.1) to (N.GT.2) and line 14 were omitted, the program would not run properly.

(d) An alternate way of writing line 15 is IF(N-1)20,30,20

(e) In this program LIST may contain either INTEGER or REAL numbers.

The I-N rules apply to subscripted variables.

Answer is (e) ●

PROGRAMMING 11

On data cards, numerical values of data that are to be input must be located

(a) between column 7 and 72
(b) between column 1 and 72
(c) between column 1 and 80
(d) between column 1 and 7
(e) between column 7 and 80

The full 80 columns may be used when punching data into cards. ●

Answer is (c)

PROGRAMMING 12

Select the incorrect FORTRAN statement.

(a) `WRITE(61,10) ((I(I,J),I=1,10),J=1,10)`
(b) `WRITE(61,10) (A(I),B(I),I=1,10)`
(c) `WRITE(61,10) A,B,C,(D(I),I=1,10)`
(d) `WRITE(61,10) (A(I),I=1,10),(B(I),I=5,10)`
(e) `WRITE(61,10) I,K,(D(J),J=5,10)`

I(I,J) is incorrect. ●

Answer is (a)

PROGRAMMING 13

Select the incorrect FORTRAN statement, if there is one.

(a) `X = FLOAT(I)*Y+Z`
(b) `J = INT(Z)+K`
(c) `R = FLOAT(K)+FLOAT(L)+M`
(d) `A = (A+B)+(C+D)`
(e) All are correct.

Item (d) contains parentheses that do not affect the computation. Thus they are meaningless, but not incorrect. The mixed mode arithmetic in Item (c) is correct FORTRAN IV syntax.

Answer is (e) ●

PROGRAMMING 14

Given: DIMENSION A(3,2,6), B(20)

Which one of the following is correct FORTRAN to EQUIVALENCE A(2,1,4) and B(6)?

(a) EQUIVALENCE (A,B)

(b) EQUIVALENCE (A(1),B(1))

(c) EQUIVALENCE (A(20),B(6))

(d) EQUIVALENCE (A(1,1,1),B(1))

(e) EQUIVALENCE (A(2,1,4),B(6))

An EQUIVALENCE declaration allows different names to be used for the same storage location in memory. This might be done to reduce the amount of memory required, or for ease in programming. Here the equivalence is to be of a singly subscripted variable and a triply subscripted variable.

The EQUIVALENCE declaration must be written with single subscripted notation irrespective of whether the variables are singly, doubly, or triply subscripted. [The exception would be that item (a) above is an acceptable substitute for item (b).]

Arrays are stored in memory in consecutive sequence. For A(3,2,6) the order of storage is

A(1,1,1), A(2,1,1), A(3,1,1), A(1,2,1), A(2,2,1), A(3,2,1)

A(1,1,2), A(2,1,2), A(3,1,2), A(1,2,2), A(2,2,2), A(3,2,2)

A(1,1,3), A(2,1,3), A(3,1,3), A(1,2,3), A(2,2,3), A(3,2,3)

A(1,1,4), A(2,1,4) By counting one finds that A(2,1,4) is the 20th storage location. This is to correspond with the 6th storage location for B.

A proper EQUIVALENCE declaration is EQUIVALENCE (A(20),B(6)).
Note that equally correct would be any linkage that aligns the two arrays in the desired manner like EQUIVALENCE (A(19),B(5))
or EQUIVALENCE (A(15),B(1))

Answer is (c) ●

PROGRAMMING 15

The binary representation **10101** corresponds to which base 10 number?

(a) 3
(b) 16
(c) 21
(d) 42
(e) 10,101

The binary (base 2) number system representation is

	2^4	2^3	2^2	2^1	2^0
	16	8	4	2	1
binary number	1	0	1	0	1
	16	0	4	0	1 = 21 ●

Answer is (c)

PROGRAMMING 16

Consider the following portion of a FORTRAN program

```
      X = 0
      DO 10 I=2,20,4
      DO 10 J=2,20
   10 X = X+1
```

After these statements have been executed, the value of X is

(a) 24
(b) 95
(c) 361
(d) 400
(e) Not Listed

The outer DO is executed for values of I=2,6,10,14,18 = 5 times.
The inner DO is executed for values of J=2 through 20 = 19 times.

$19 \times 5 = 95$ ●

Answer is (b)

PROGRAMMING 17

In BASIC, which one of the following is untrue?

(a) Every statement normally is placed on a separate line.

(b) Each statement must have a line number.

(c) Blank spaces may be inserted where desired to improve readability of the statement.

(d) Successive statements need not have increasing line numbers.

(e) Each statement must have a BASIC keyword indicating the type of instruction.

Normally successive statements in BASIC *must* have increasing line numbers.

Answer is (d) ●

PROGRAMMING 18

Which one of the following is *not* a BASIC statement?

(a) `IF A>B THEN 60`

(b) `FORMAT (5X,F10.1)`

(c) `REM SUM`

(d) `INPUT A,B`

(e) `LET X=1`

Item (b) is a FORTRAN Format statement for reading or writing a floating point number. In BASIC, printing of variables is done using a PRINT statement.

Answer is (b) ●

PROGRAMMING 19

A loop in BASIC begins with a FOR-TO statement. Which statement closes this loop?

(a) `NEXT`

(b) `GOTO`

(c) `CONTINUE`

(d) `STOP`

(e) `END`

Answer is (a) ●

Index

Acceleration, normal, 39-41
radial, 37-41
tangential, 39-41
translational, 47
Adiabatic process, 165
saturation temp., 169-170
Air conditioning, 149
Amphoteric hydroxide, 239
Annual cost, 213
Annuity, defined, 221
Arithmetic mean, 54
Atomic number, 245-246
Austenite, 93, 96-97
Average annual cost, 228-229
Average interest, 229
Avogadro number, 236

Bandwidth, 205-206
BASIC, 262
Beam, bent, 80-81, 85-86
cantilever, 59, 81-82
fixed ends, 84
statically indeterminate, 67-68
Beam, bending moment, 76-77
deflection, 56-57, 59-60, 64, 68, 75-76
reactions, 79
stress, 73-74, 82
Bernoulli equation, 119, 134
Binding energy, 241
Bionomial, 2
Boiling point, 237
Brayton cycle, 160
Buffer solutions, 240
Bulk modulus, 249
Buoyancy, 106, 109-113

Capitalized cost, 212, 231
Carnot cycle, 150, 152, 154, 158-159
Cavitation, 130
Cement, 239
Center of gravity, 107
of pressure, 107
Centroid, 27-28, 32
Chemical equilibria, 240
formula, 235-236
reactions, 235, 240
Chemistry, 235-240
Circuit, bridge, 191
network, 193
Circuit, series; parallel, 188-189, 195-196
steady state, 192, 202
transients, 192
Coefficient of coupling, 199
of Performance, 151-152
of restitution, 35
of variation, 54
Column, 52, 66, 71
Combinations, 2
Combustion, 173-178
Complex conjugate, 3
numbers, 3, 11, 186-187, 194
exponential form, 194
polar form, 194
trigonometric form, 194
Composite structure, 50, 57-58, 69
Compound amount, 209-210
interest notation, 209-210
Compressibility, 137
Computer programming, 253-262
Concrete, 50, 99-101
Conduction band, 91
Conservation of energy, 116
Conic sections, 6, 9
Cosines, law of, 3
Continuity, 113-114, 116
Coulombic attraction, 195
Critical depth, 119
Cross section, 247
Crystal imperfections, 89
Curve symmetry, 7

Darcy-Weisbach equation, 125-126, 134
Decay constant, 244-245
Derivative, 1-3, 7-8, 11, 13-14
partial 12
Determinant, 14
Dew point, 169, 177
Diffraction, 248-249
Dimensional analysis, 87, 124
Diode rectifier, 200
Displacement, 35-37, 44-45
Distance, between points, 5
Distance-velocity-acceleration, 44-45, 47
Doppler effect, 250
Dynamic force, 120
Dynamics, 33-48

Economic analysis, 207-234
Effective interest rate, 218

Efficiency, 194, 201-202
    carnot cycle, 150-151, 159
    mechanical, 36
    motor, 201
    thermal, 150
    transformer, 201-202
Elastic range, 53
Electrical, circuits, 187-206
    resistivity, 236
Electrolysis, Faraday, 195
Electron, orbitals, 241
    flow, 195
Elongation, 65
Energy flow, 115-116, 156, 172
    kinetic, 115, 118, 143, 156
    potential, 115
    pressure, 115-116
    spring, 34, 53
    total, 118-119
Energy balance, 117, 122, 125-126, 148
    gradient, 119
Engineering economy, 207-234
Enthalpy, 138, 140, 147-148, 156 166
Entropy, 138-139, 153-155
Equation, roots, 5
    of state, 137
Equilibrium, 18, 22, 25-29, 161
Equivalent circuit, 202
    weight, 237
Euler equation, 52, 71
Eutectic, 94-97
Exponents, 10

Factorial, 2, 8, 10
Fatigue failure, 98-99
Ferromagnetic, 93
First law of thermodynamics, 119, 140-142
Flexure formula, 74
Flow, laminar, 122
    turbulent, 122-123
    nets, 123
Fluid mechanics, 103-136
Force, centrifugal, 42
    gravitational, 42, 46
    radial, 34, 37
    reaction, 34
    spring, 34, 53
Force-mass-acceleration, 35, 43-44
FORTRAN, 252-261
Frame, bent, 85-86
Frequency, 248, 250-251
Friction, 18, 37, 109
    factor, 124-125
Future worth, 213

Gas mixtures, 167-168
    solubility, 239
Geometric mean, 206
Grain size, growth, 89, 97
Gradient factor, 228

Half-life, 244
Half power points, 205-206
Halogens, 238
Head, static, 105
    loss, 117, 119, 122, 124-125, 127
Heat, 138, 142
    balance, 143-147
    capacity, 140-141, 149, 168
    exchange, 146, 186
    pump, 151-152, 160
    transfer, 141, 143, 187-189
        conduction, 179-183, 187
        convection, 195
        film coefficient, 183-184
        overall coefficient, 182-186
        radiation, 186-188
Heat of vaporization, 139
Hooke's law, 52, 71
Hydraulic gradient, 119
    head, 118
    jump, 133
    radius, 132
Hydrogen ion concentration, 140

Ideal gas law, 137, 167, 176
Impact, oblique, 35
Impulse-momentum, 43, 46, 120-122
Inductance, 200
    mutual, 199
Infinite life, 212, 224
Inflection point, 1-2
Integration, 1, 13
    graphic, 36
Interferance, 248-249
Intermetallic compounds, 94-95
Internal energy, 141-142, 155-156
Iron-iron carbide phases, 96
Isentropic process, 153, 162, 164-165
Isothermal process, 166
Isotopes, 241

LeChatelier's principle, 240
Light intensity, 252
Limits, 8
Load, allowable, 50
Logarithms, 7, 10-11

Mach number, 170-171
Martensite, 97-98
Mass flow rate, 114, 116, 143, 147, 171-172
    number, 241, 245-246
Materials science, 87-102
Mathematical modeling, 15-16
Mathematics, 1-16
Matrix notation, 14
Maximum, 1-2
Mechanics of materials, 49-86

Metal, annealing, 90-91, 97
 coldworking, 90-91
 hardening, 90-91
 heat capacity, 238
  treatments, 91, 97
Meter scale change, 194
Minimum, 1, 2
Modulus, elastic, 87-88, 249
 of resilience, 88
 of rigidity, 71
 shear, 50-51, 71, 249
Mohr's circle, 58
Mole, 238
 fraction, 167-169, 176
Mollier diagram, 138
Moment, 17, 61-63
 bending, 70, 72
 diagram, 61, 63, 80
 of inertia, area, 19, 21, 29, 59, 107
 mass, 23
 polar, 49
 of momentum, 41
Moody diagram, 124-125, 134
Motion, curvilinear, 39
Motor, shunt, 201

Network analysis, 197-199, 203
 equations, Kirchhoff, 193, 195
Neutrons, 241, 246-247
Nominal interest rate, 210
Nuclear particles, 242
 reactions, 245-247
Number series, 4

Open channel flow, 133-136
Orifice coefficient, 128-129, 132
Oxidation-reduction, 235

Parallel forces, 31
Partial pressure, 167
Pearlite, 96-98
Period, of pendulum, 48
pH, 240
Phase rule, 137
 reactions, 96, 98
Photon, 242, 251
Pipe networks, 128, 134
Pitot tube, 131, 171-172
Plastic deformation, 90
Plastics, 102
Poisson's ratio, 51, 70, 85
Polytropic process, 162-3
Power, 36, 143, 157, 187-188
 pumping, 117
 reactive, 187-188
Power factor, 187-188, 190
 correction, 187-188, 204
Present worth, 211

Pressure, atmospheric, 105
 stagnation, 119, 131
 static, 66, 104-105, 107-108, 111
Pressure vessel, 60, 77-78, 110
Probability, 2, 9, 13
 conditional, 9
Programming, computer, 253-262
Projectile path, 48
Protons, 241, 246, 251
Psychrometric chart, 169
pump, centrifugal, 131

Quality factor, 206
Quantum numbers, 242
Quasistatic process, 153, 162

Radioactive decay, 243-244
Radio carbon dating, 245
Radius of gyration, 46, 52
Rate of return, 214
RC time constant, 196-197
Reactance, 190
Real gases, 166
Recyrstallization temperature, 90
Reflection, 248-249, 252
Refraction, 248-249, 252
Refrigeration, 185-186
Relative humidity, 169-170, 176
Relative motion, 33, 130
Resistance, temperature coefficient 191-192, 203
Resistivity, electrical, 200, 203
Resonance, 205-206
Reynolds number, 122-125
rms values, 204-205
Roughness ratio, 125
Runoff, rainfall, 135
Rusting, 99

Saturated liquid, vapor, 139
Second law of thermodynamics, 142
Section modulus, 59
Sections, 17
Semiconduction, 92
Shaft work, 155-156, 162
Shear diagram, 61, 63, 80, 85
 modulus, 50-51, 71
Sine curve, 3
Sines, law of, 39
Sinking fund, 210
Siphon, 130
Slenderness ratio, 52, 71
Solid solutions, 93-94
Sonic velocity, 87, 171
Sound, 249
 intensity, 249
Space frame, 19-21, 31
Spontaneous processes, 154
Spring constant, 34, 53

Stability, 24, 26-27
Stagnation properties, 170, 172
Standard deviation, 9, 54
Static indeterminacy, 65, 68
Statics, 17-32
Steam tables, 148, 164
Steel, phase transformations, 97-98
    stress-strain, 88
Steels, 95
Stefan-Boltzmann constant, 188
Straight line depreciation, 216
Strain, allowable, 50
    energy, 89-90
Stress, allowable, 50
    concentration, 70
    hoop, 66
    principal, 58
    rupture, 98
    shear, 64
Stress-strain, 52-53
Structure, composite, 50, 57-58, 69
Structure of matter, 241-252
Support reaction, 17, 25-27, 55

Temperature scales, 139
Thermal conductivity, 179-180, 183, 236
    expansion, 51-52, 63-64, 71, 73, 83, 85
    resistance, 180, 183, 188
Thermodynamic cycles, 157-158
    properties, 138
    processes, 159, 161-162
Thermodynamic state, 137, 139-140
Thermodynamics, 137-186
Thermonuclear reaction, 146
Throttling, 145
Time domain, 204-205
Torque, 43-44, 48-50
Torsion, 49-51, 64
Torsion formula, 49, 51
Transformations, Δ-Y, 189-190
Transformer, 201-202
Transistor, 92, 187-188
    bias, 92, 188
    carriers, 91
    circuits, 187-188
Trigonometric functions, 4-6, 11-12
Truss analysis, 17, 25-26

Valence band, 91
Vector addition, 23, 40
    quantity, 40
Velocity, angular, 37
    normal, 39-41
    radial, 39-41
    relative, 35
    tangential, 33-34, 38-41
    translational, 45
Velocity-time plot, 36
Venturi, 116
Viscosity, 103-104
Virtual work, 67
Voltage measurement, 196
Volume flow rate, 113, 116

Wave, electromagnetic, 248-249, 251
    energy, 248, 251
    longitudinal, 249
    transverse, 248-249
    velocity, 248-252
Wavelength, 250-251
Water hardness, 237
Welds, 74-75, 78-79
Work, 138, 141, 155
    closed system, 156, 163
    open system, 162
    pumping, 119